Arnulfo L. Oliveira Memorial Library

# The Producer as Composer

# The Producer as Composer

## Shaping the Sounds of Popular Music

Virgil Moorefield

The MIT Press
Cambridge, Massachusetts
London, England

MIT Press books may be purchased at special quantity discounts for business or sales promotional use. For information, please email special_sales@mitpress.mit.edu or write to Special Sales Department, The MIT Press, 55 Hayward Street, Cambridge, MA 02142.

This book was set in Stone Serif by Binghamton Valley Composition LLC, in Quark Xpress, and was printed and bound in the United States of America.

Library of Congress Cataloging-in-Publication Data

Moorefield, Virgil, 1956– .
  The producer as composer : shaping the sounds of popular music / Virgil Moorefield.
    p.  cm.
  Includes discography (p.   ) and bibliographical references (p.   ) and index.
  Contents: From mirror to beacon—The studio as musical instrument—The producer takes center stage.
  ISBN 0-262-13457-8 (hc : alk. paper)
  1. Popular music—Production and direction—History and analysis. 2. Sound recordings—Production and direction—History and analysis. I. Title.

  ML3470.M66  2005
  781.49—dc22

                                                                      2005041588

10 9 8 7 6 5 4 3 2 1

In memory of my parents,

Virgil Hisgen Moorefield
and
Jane Moorefield Heumann

The bass is so low you can't get under it
The high is so high you can't get over it
So in other words be with it.

—Kool Herc

# Contents

# Acknowledgments

I would like to acknowledge the encouragement and support of Paul Lansky, who was my advisor at Princeton.

Thanks also to Barbara White, who read early versions of my work closely and provided important suggestions, and to my research assistants at Northwestern: Charles Larish, Michael Kraskin, John Woodruff, Mark Cartwright, Jeffrey Weeter.

I had many interesting discussions on the subject of the producer in pop with my wife, Emily XYZ, who is a walking encyclopedia of rock and pop history.

Thanks to Gary Kendall for his unwavering support.

Others who helped include David First, Michael Hopkins, Joshua Fried, Perry Cook, Tom and all the knowledgeable folks at the Princeton Record Exchange, and my students at Princeton and Northwestern.

My thanks also to Robert Rowe, and to Doug Sery and the production team at MIT Press.

# Introduction

Over the last fifty years, the philosophy and technique of music production have undergone a major transformation. As the activity of recording has widened in scope from a primarily technical matter to a conceptual and artistic one as well, it has assumed a central role in areas such as instrumental arrangement and the sculpting and placement of audio samples. The concept of a *sound* in the sense of stylistic choice, and the ability to capture and mold it, have grown in importance as recording technology has become increasingly complex. Contemporary conceptions of the role of the producer have been broadened through the example set by the work of a number of extraordinary figures such as Phil Spector, George Martin, and Brian Eno.

Each of these people expanded the purview of the modern producer by introducing radical new concepts such as the Wall of Sound (Spector), the confluence of classical, commercial, and experimental techniques (Martin), and the studio as a musical instrument in its own right (Eno). Factors such as technological development, the ascendancy of the recorded work over live performance, and audience tastes have also contributed to the development of the producer's role in the studio.

In this book, I make the case for three central developments in production and claim that they are all driven by an underlying mechanism. One: recording has gone from being primarily a technical to an artistic matter. Two: recording's metaphor has shifted from one of the "illusion of reality" (mimetic space) to the "reality of illusion" (a virtual world in which everything is possible). Three: the contemporary producer is an *auteur*. The underlying mechanism is technological development, encompassing both invention and dissemination due to economies of scale.

While many approaches to production are discussed, my interest ultimately lies in the producer as composer. There are many different kinds of producers working today; the development of production methodologies has been more additive than evolutionary. But as Evan Eisenberg has pointed out, "it is the artist-producer, the musical creator whose impulse is to create records, who plays the central part in the development of phonography as an art" (Eisenberg, 128). There is always a collaborative aspect to working in the studio, and the importance of studio musicians, recording engineers, and other contributors to production should always be acknowledged. Yet contemporary music such as hip-hop is conflating the role of the producer and that of the auteur, notably at the time of this writing in the form of the number one album on the pop charts, Outkast's *The Love Below/Speakerboxx*, which was produced by the artists.

Originally, the aim of recordings was to create the illusion of a concert hall setting. The idea was to bring to the living room the sensation of being at a live performance—a metaphor of presence. This was true of all types of music, until rock and pop took a different turn, and became the principal area of creative expression for the producer. To understand why this is so, one might consider what some of the fundamental aspects of different genres of music are, and how these influence their recording aesthetics.

In classical music, faithful interpretation of the written score has long been central. Western notation predates recording by centuries; it developed into a very precise set of instructions for performance. From the point of view of recording, notation is an attempt to describe a piece in such detail as to make it completely repeatable. Jazz is based on a combination of written arrangements and improvisation in the form of personal, real-time interpretations of a canon of "heads." Both jazz and classical music are played by highly trained performers, who have spent a lifetime perfecting their ability to read music and, in the case of jazz, to improvise in a certain way. Thus these forms of music are generally recorded live, without overdubs, and with minimal, if any, postperformance enhancement. The idea is to capture the live performance in the tradition of "realistic" recording, as it has existed since the 1870s.

Rock and the many subgenres it has spawned are a different story: timbre and rhythm are arguably the most important aspects of this music. Generally, nothing beyond a lyric sheet and possibly a few chord changes is written down; the recording of a song functions as its score, its definitive version. It is no accident that the rise of rock'n'roll happened almost at the same time as fundamental technological innovations such as tape editing and overdubbing. For rock and pop, the interest generally lies not in virtuosity or harmonic complexity, but in a mood, an atmosphere, an unusual combination of sounds; these are greatly enhanced by good production.

Although rock recording did begin as a largely mimetic enterprise, the matrix which underlies pop recording was reversed in the mid-sixties. There came into being a new conception of making records, developed separately and in stages, most notably by Phil Spector and George Martin. While different in many ways, both of their approaches to production involved replacing the quest for the ability to present the illusion of physical reality with a new aesthetic. The new sonic world they sought to create was the appearance of a reality which could not actually exist—a pseudo-reality, created in synthetic space.

Spector's Wall of Sound and Martin's eclectic arrangements with the Beatles were clearly breaks with the aesthetic of realism which had been prevalent until that time. The amassing of huge numbers of instruments awash in cavernous ambience, the juxtaposition of tape loops with string quartets, the manipulation of figure and ground through placement in the mix were all manifestations of a new aesthetic. Their interest lay not in replicating the natural world, but rather in transforming it into something else: by embracing subjectivity, they made a potent argument for viewing the producer as auteur. In the context of the then very limited availability of professional-grade recording technology and techniques, they were in effect attaining a status akin to that of a film director. The *mise-en-scène* which had always been the domain of the record producer widened in scope, as did his artistic role.

The work of these pioneers left a lasting mark on the industry. In addition to traditional responsibilities as talent scout, organizer, and general overseer of the recording project, the active involvement of the modern producer often encompasses areas such as style, arrangement, and musical form, elements formerly associated exclusively with arrangers and composers. In many cases bandmembers may become actors in the producer's "movie for your ears," as Frank Zappa referred to the Mothers of Invention's double album *Uncle Meat* in its liner notes. Producers who are also composers, such as Brian Eno, Bill Laswell, and Trent Reznor, generally put their imprint on every project they are associated with. Eno's "Oblique Strategies," usually involving some form of chance operation, Laswell's

stable of world music instrumentalists, Reznor's trademark electronic distortions: all are omnipresent on the records they have made, be it under
their own name or as producer of someone else's project. Today there are
numerous name-brand producers who function more as active agents than
as catalysts.

This situation of the producer reaching into the actual stylistic direction
and sound of the recorded artifact is the result of a number of factors
which have combined to enhance his importance and status. First, there is
the matter of expertise. Until very recently, the equipment necessary to
produce professional-grade recordings was prohibitively expensive, and
the knowledge required to obtain good results the province of a few highly
specialized technicians. As the mediator between the two worlds of inspi-
ration and know-how, the producer became a central figure. Second, there
is the primacy of recordings over live performance. Today far more people
listen to recorded music than to live musicians; hence the sound of the
recording is what is most important, and contributes heavily to making or
breaking an act. Third, there is the culture's appetite for a constant parade
of young faces and new styles (mostly of the visual variety), which stands
in contrast to the practical need for someone with experience. The image
of fashion models on the runway comes to mind, while the forty- or fifty-
something designers oversee their crews backstage to ensure a successful
presentation. The balance between the elements is more fluid in pop
music, especially in recent times where the producer and the artist are
often one; but there have been instances of the producer totally determin-
ing the sound of the act, as for example in the case of the English band Joy
Division.

In a general sense, these developments have caused a shift in musical
values toward production. As the cultural ear becomes more familiar with
what is possible, it is able to distinguish between different levels and styles
of production, and becomes more aware of the production aspect of a
track.

Shaping the sound of a modern pop record usually involves a division of
labor. There have probably been as many varieties of working arrangements as there have been producers: in the case of early Motown, there was
one executive producer (Berry Gordy) who had final say on every aspect of
a release, while a staff of engineers, producer–songwriters, and musicians
came up with the product. In such situations, it is difficult to say exactly

who does what. Indeed, there are many different conceptions of what a producer does. What is clear is that the producer who puts his name on the record takes responsibility for the overall production. As Albin Zak points out, "while no two producers have quite the same combination of skills, each must have the ability to draw together diverse elements and to manage the dynamics of collaborative creativity among the members of the recording team" (Zak, 173).

Often the producer will consult with the engineer in order to determine how best to achieve the desired sonic effect. Far more than his counterparts in other types of recording such as jazz or classical, the contemporary rock and pop producer is usually well versed in technical concepts such as the subtleties of ambience, compression, distortion, and effects such as phase-shifting, harmonizers, and so on. In addition, he must either have the requisite skills to achieve the sonic output he desires, or have a close working relationship with a highly skilled engineer—preferably both. In any event, the producer almost always acts as some sort of orchestrator, be it in the choice of electronic timbral and spatial effects described above, a more traditional instrumental enhancement such as a string arrangement, or both (e.g., as with George Martin).

The creative involvement of the producer in the shaping of a record's sound also reflects how technology and artistic creation are increasingly interdependent in our culture. Just as the sound of the symphony orchestra which powered the works of the great symphonists reflects the mechanical technology of the seventeenth through nineteenth centuries, so modern pop recording has embraced the analog and digital technologies which characterize our cultural epoch. As recordings which utilize advanced technologies in new ways set trends and turn a profit, there is greater demand for new equipment, which in turn pushes the envelope of what is possible in the studio.

In the last ten years, some truly amazing technological developments have come to market, coupled with a major price drop in digital recording and signal-processing equipment. The equivalent of a recording console which cost $150,000 in 1995 can now be had for about $2,000, owing to cheap digital memory, miniaturization, and the increasingly globalized economy. This has made possible the emergence of the all-in-one figure, a sort of apotheosis of the producer as composer. Acts such as the Chemical

Brothers, DJ Shadow, the Dust Brothers, Daft Punk, Squarepusher, and others have dispensed with the traditional separation between artist, engineer, and producer altogether, making themselves the embodiment of all three. There has been a return to the idea of pure electronic music as first implemented by Vladimir Ussachevsky, Pierre Schaefer, and others in the fifties, but now in popular music with certain stylistic conventions such as loops and break-beats: a sort of cross between the image of the scientist in his labcoat at the mainframe with the DJ adding effects devices to his turntable rig.

The robotic, 16th-note-grid drum machines of the seventies influenced the gestures of early rap much as the new generation of producer–composers are influenced by the machines they use. Because their only product is electronic and their available sources potentially consist of the sum total of all recordings ever made, all sound sources become equal: on this level playing field, it doesn't matter whether producers are the original creators of the sounds they work with or whether they slice and dice the recorded work of others. Thanks to the integration of digital control and digital audio in the computer, an unprecedented amount of sonic control is available to everyone—there exists today unparalleled creative opportunity for the individual of even moderate technical ability. The availability of inexpensive resources for sonic manipulation will continue to radically alter the sound of pop music in the future.

I have endeavored to outline the historical development of the role of the producer, and to show how the job description gradually became all-encompassing, to the point where producers are now taking to the stage themselves and utilizing portable recording studios as an electronic orchestra.

Most of the music I discuss was created in the last thirty-five years. On one hand, this is good, because many people remember a lot about the music itself and the circumstances under which it was created. On the other hand, almost nothing has been written about pop music in the kind of detail which is available about classical music. There is therefore hardly any serious, detailed writing about what is actually heard in any given pop song. Yet this is precisely what's so interesting about good pop: why does it sound the way it does? What went into its creation? In answering these questions, the role of the producer becomes clear; or at least, the musical issues facing artists, engineers, and producers are raised.

In this book, "pop" means the lighter side of popular music, such as Motown, Michael Jackson, and the Beach Boys. "Rock" denotes the harder sound associated with Hendrix, the Who, and the Rolling Stones.

I have used sources when available. Although most writing about pop is prone to anecdotes and short on musical detail, it is sometimes surprising what one can find in a biography of a famous producer or band. I've also relied on my twenty years of recording studio experience (I owned and operated a recording studio in New York City for thirteen years, and recorded as a drummer and composer in a few others) in order to interpret what one hears on the recordings talked about in this essay. In doubtful cases, I bent the ear of many a knowledgeable friend in order to arrive at what seemed a reasonable conclusion. In a couple of cases, I was present at the recording sessions themselves.

It is highly recommended to listen to the music I describe in order to follow the discussion. For the most part, the tracks are readily available in any good record store, or from an online music service.

# 1    From Mirror to Beacon

## Beginnings

In the early days of recording, the record producer in the modern sense did not exist. Thomas Edison had invented the phonograph in 1877, and by 1887 Emile Berliner had devised the gramophone. Perhaps the closest thing to production in the 1890s is the image of Fred Gaisberg, who ran the first recording studio, holding an opera singer by the arm and moving her closer or further away from the gramophone's horn according to the dynamics of the passage being sung (Gronow and Saunio, 8).

These were radically new machines, and the gramophone would have a huge impact on music as a whole; nevertheless, recording itself was initially a rather cut-and-dried affair. Because of the limited frequency range of recording devices and their relative lack of sensitivity and dynamic range, as well as the popularity of opera at the time, the first recordings were mostly of operatic arias. The strong voices of opera singers recorded relatively well, as attested to by the fact that Enrico Caruso's recordings from the first two decades of the twentieth century have remained in print continually to this day.

At first, recordings had to be made practically by hand. There was no electricity involved; a bullhorn picked up sound, and the attached stylus etched grooves into a roll, or a disc. Since there was no way to duplicate records, early recording engineers would have to line up ten or so phonographs

in front of a loud-voiced singer or a small brass band, and in this way ten recordings could be made of one performance. . . . When the Irish-American comedian, Dan Quinn, told Phonoscope in 1896 that he had made 15,000 recordings in the past month, he had, in fact, had to sing his favorite songs hundreds of times to put them onto wax. (Ibid., 4)

With the adoption of the microphone, which had been invented for telephony in 1878, recording fidelity increased dramatically. The first known music recording experiments using a microphone were made in England in 1919, and with the changeover from mechanical to electrical recording, which took place roughly from 1920 to 1925, recording engineers started to have more choices about how to shape the sound of recordings.

Given the current discussion about whether analog or digital sounds "warmer," it's interesting to take a look back and discover that at one time, the debate was as to whether mechanical or electrical recording was preferable:

Fred Gaisberg long ago noted that "the velvet tone of Kreisler's violin, for some unknown reason, was best in those old records and has never been recaptured by the electrical process." He added that "in some ways acoustic [i.e., mechanical, as opposed to electrical] recording flattered the voice. A glance at the rich catalogue of that period will show that it was the heyday of the singer." [ . . . ] In 1930 Mackenzie wrote of the new microphone-made records, "I do not believe that any audience could sit still and listen nowadays to hours of electrical recording and remain sane." (Eisenberg, 112)

When the legendary producer John Hammond began his career in the early thirties, recording classic artists such as Bessie Smith, Billie Holiday, and Benny Goodman,

the producer's duties in the studio were fairly simple. The most important job was to contact the artists and agree on the numbers to be recorded. If the band was not a permanent one, he would procure the accompanists and arrangements. A good record producer was, above all, a talent scout. In the studio the music was recorded as "naturally" as possible, as in a live performance. With the technology available it could not be altered appreciably; there were a few microphones in the studio. The recording was made by etching the performance straight onto a wax disc, and if it did not succeed the first time, the whole thing was done again. (Gronow and Saunio, 70)

John Hammond was a remarkable figure in the history of record production. His involvement with studio recording spanned several decades, genres, and styles. Hammond was also a jazz critic (as was Voyle Gilmour, an early producer for Frank Sinatra). This is not surprising, because the role of the producer in the studio at that time was to be a critic, and, if he was really doing his job well, to help bring out the best in a performer.

The aesthetic of the early period of record production (roughly pre-1950) is summed up by another famous producer of that era, Mitch Miller, who is best known for his hit television series of the sixties, *Sing Along with Mitch*. Asked how he managed to switch back and forth between producing classical recordings and popular acts such as Frankie Laine and Vic Damone, Miller replied:

I'm always surprised when people say it's a switch. I never compartmentalized it in my own mind. The same rules apply. You know—taste, musicianship, balance, get the best out the artist. Many times the artist doesn't know what his best characteristics are, and you're there to remind them. You can't put in what isn't there, but you can remind them of what they have and they're not using. (Quoted in Fox, 34)

Like Hammond, Miller was ambivalent at best about multitrack recording, faulting the modern processes of overdubbing and punching in as techniques which rob music of its spontaneity and vitality. For both producers, and probably for their generation as a whole, modern recording processes such as editing, splicing, overdubbing, and remixing were a form of dishonesty. In this worldview, a valid musician is a virtuoso, and the ability to perform in real time is paramount. The aesthetic which governed early recording was one in which the concert hall experience was to be recreated as faithfully as possible. Real performances were seen as a necessity. This view has disappeared only gradually as recordings have become the primary way in which listeners experience music.

In the later part of his career, John Hammond produced recordings by such major pop figures as Bob Dylan and Bruce Springsteen. He was the quintessential purist producer, one who could capture performances onto acetate disc which were the "real thing"—that is to say, unedited. Hammond was making a virtue of necessity, because it wasn't possible to edit recordings until tape came along. Reminiscing about the recording of a Mozart divertimento with conductor Max Goberman in 1937, Hammond says:

Max, who was a good leftist, didn't believe in using anything but unemployed French horn players, and two of the most important French horn parts in the history of chamber music were in that divertimento. There were bloops on practically every side, and there was no way we could edit them out. So it stayed, bloops and all, and it's still in the catalog. Tape didn't come out until after World War II. (Ibid., 5)

In fact, the German company AEG had already invented the first magnetic tape recorder in 1935. In 1945, an American G.I. named John Mullin brought home two tape recorders taken from Radio Frankfurt as spoils of war, and persuaded a company named Ampex to manufacture a few copies. The rest is history: the tape recorder was in general use by 1948.

The first important effect of the introduction of tape was the possibility of splicing. Different takes could be composited into one performance, or a flawed section of a recording could be re-recorded and edited in. Tape was still monaural, though, so true multitracking was not yet possible.

The next and possibly most significant step after the invention of recording itself was the invention of true overdubbing. The idea was around before the introduction of tape recording to the U.S. In 1941, saxophonist Sidney Bechet made one of the first multitrack recordings onto acetate disc and called it "Blues of Bechet." He was recorded playing along with a recording of himself stored on acetate disc. The process was then repeated a number of times. But because the original signal lost a generation every time an overdub was made, it was not possible to attain a high degree of fidelity. Although the recording was released commercially, the record companies regarded it and other early attempts at multitracking by artists such as Patti Page as novelties.

An eccentric guitarist and inventor named Les Paul had also been experimenting with acetate overdubs since 1930. Originally from Waukesha, Wisconsin, Les Paul moved to Los Angeles and signed as a recording artist with Capitol Records in 1947. With his partner and future wife Mary Ford, Paul recorded a number of hits and popular radio commercials, thus gaining valuable studio experience. A tinkerer by nature, Les Paul did his own recording in his own studio, and got involved in the technical details of recording his music. The results were of such quality that "his ten-inch album, *New Sound,* was seized upon by hi-fi manufacturers and retailers as a test disc with which to demonstrate the quality of their gramophone products" (Cunningham, 28).

Les Paul is probably best known for the electrification of the guitar; the mellow-toned Les Paul model electric guitar has been manufactured by Gibson ever since the fifties, and it is still quite popular today. Yet his experiments with various placements of recording and playback heads on the tape recorder is at least an equally important contribution. Paul realized that if another recording head were placed next to the one already there, two signals could be recorded in sync with each other, without generational loss, separately or at the same time. In addition, when a playback head was placed behind the record head on a tape machine, signals could be played back out of sync, resulting in now-standard effects such as phasing, flanging, chorus, and delay. Although their names may suggest otherwise, these effects can all be obtained by simply delaying a signal against itself at various tape speeds, and have become staples of modern pop recording.

Paul wasted no time in applying these discoveries to his own music, and he had a number one hit in 1951 together with Mary Ford called "How High the Moon." Les Paul had let the genie of multitracking out of the box, and although the recording industry wouldn't realize it for a few years, there was no turning back.

## The Brill Building Songwriters

In 1958, Al Nevins and Don Kirshner formed Aldon Records and set up shop in New York City at 1619 Broadway, also known as the Brill Building. They soon found themselves running a hit factory, employing songwriting teams such as Burt Bacharach and Hal David, Carole King and Gerry Goffin, as well as Jeff Barry and Ellie Greenwich. They were responsible for many of the greatest hits in American pop music from 1959 to 1964. There was a ready market for the work produced by these songwriting teams: rock'n'roll had exploded across America, and record companies such as Columbia, RCA, Atlantic, and ABC needed a large quantity of songs for their artists. At Aldon, writers were cooped up in rows of small cubicles with pianos, feverishly churning out songs on a full-time basis. The result of their work was usually a piano and voice arrangement, known as a lead sheet.

Working independently out of their office on nearby West 57th Street, songwriters Jerry Leiber and Mike Stoller were producing hit after hit for Atlantic Records. But for Leiber and Stoller (best known for their work with Elvis Presley, who recorded songs of theirs such as "Hound Dog" and "Jailhouse Rock"), the lead sheet had proven to be too vague and open to often misguided interpretation. As Mike Stoller says,

When we conceived these things, we conceived them as records rather than as a song that exists on a piece of paper—it wasn't music that was written down in a full scale arrangement, though, when we started, so the ideas we had at the beginning, when we took our songs to an A&R man, never came back in the way we had imagined them, so we became producers in self defense. (Quoted in Tobler and Grundy, 11)

Their dissatisfaction was to have significant consequences. The songwriters became producers of their own material as well as the work of others, and they also started tutoring a number of Brill Building writers in the art of record production.

Before Leiber and Stoller stepped into the studio and became, along with Les Paul, some of the first producer–songwriters—writers who produced (directed) the recording of their own musical material in the studio— things were quite different. The usual practice at the time was that

there were A&R men who worked for RCA, CBS, MGM, and Capitol, and for the most part, they were rather astute, usually musicians or arrangers, conductors, whose job it was to select material. Publishers had regular meetings with these A&R men, and would submit material for the artists on their rosters [ . . . ]. Then the A&R man would select the material, show it to an artist, select an arranger, and the arranger would select a contractor to hire musicians. The song would be arranged, and the artists and the musicians would then go into the studio and make the record. (Ibid., 15)

Given such circumstances, it is easy to see why Leiber and Stoller would choose to become producers "in self-defense." Their world was not one of "Mickey Mouse swing records" (ibid., 21); theirs was not the emerging sub-urban way of life of postwar America. Their unconventional lifestyle and exposure to different styles of live and recorded music meant that they heard things differently than did the other professionals they worked with on the New York music scene. They were working in a new form called rock'n'roll, and they cultivated the unconventional lifestyle which was reflected in their music. They drew on black American culture: they hung out in black clubs, and they had black girlfriends. According to Leiber, "we thought we were black" (ibid., 22). The way they describe their style of working reveals that they had not just abstract melodic and harmonic ideas in mind, but rather a very specific set of sounds or "moves."

Leiber:   We used to just use shorthand after a while, sort of make signs. I'd say, "More Fats" [Domino] or "More Richard" [Little Richard] or "More Amos" [Milburn], "More Charles" [Ray Charles]. All these were signals for different styles pianistically. If I was talking about Toussaint [Allen Toussaint], it meant New Orleans.

Stoller:   If he said Fats, it generally meant triplets.

Leiber:   Hard triplets, at a certain tempo. (Fox, 161)

This technique is a kind of sonic collage, and foreshadows the sampling techniques which were to become commonplace as of the mid-eighties: a new song is assembled from reusing the hooks and riffs of past hits. (In the analog epoch under discussion here, they were reperformed, and reinter-preted by a musician; in contemporary sampling practice [e.g., Public

Enemy], the quote is literal—a soundbite is lifted from an existing record-ing—although here too a reinterpretation can be created by sonic transfor-mation and juxtaposition.) Leiber and Stoller were consciously assembling their songs by drawing on specific sounds they were familiar with from lis-tening to the work of their contemporaries. They were appropriating riffs they had heard on other records and assembling them into new material, adding new ideas and recycling old ones as they went along. The riffs they took and reshaped or quoted were not only about the notes, but *how* the notes were played, what the feel was—they were endowed with a particular flavor by virtue of how they had been performed by the artist in question. And the way they were heard was through recordings.

In order to put across their sonic ideas, it was in turn necessary for Leiber and Stoller to be in the studio, because what was being made was a record, and the combined elements of microphone placement, EQ, amp settings, mix, and of the performances themselves—tempo, groove or feel, dynam-ics, and so on—added up to shaping the work as a whole. Leiber and Stoller broke new ground in pop studio practice by becoming involved in the recording process to a then unheard-of extent. Where previous productions had at best been overseen by what was in effect a record company supervisor, here were two guys who had actually written the material they were in the studio with, and they had definite ideas about how things should be done.

One of their first sessions was with the legendary singer Big Mama Thornton. Their 1952 recording of "Hound Dog" later became a smash hit for Elvis Presley (albeit with cleaned-up lyrics). Leiber and Stoller's involve-ment in the recording session itself came about somewhat by chance. The person who had played drums at the rehearsals for the recording of "Hound Dog," Johnny Otis, was both a drummer and a bandleader, and was in charge of the recording session. According to Mike Stoller:

[Otis's] drummer wasn't getting the right sound, so Jerry [Leiber] ran into the booth and told Johnny to get out and play the drums, because that was the sound that was needed. Jerry stayed in the booth, I stayed on the floor in the studio, and we com-municated over the talkback, and that was how the record was made. (Quoted in Tobler and Grundy, 11)

From these simple beginnings, Leiber and Stoller developed an integral method of making records which involved composing songs at the piano, selecting artists to perform them (in some cases putting together an entire group of singers and musicians), working with an arranger and rehearsing with the musicians, and finally, supervising the actual recording sessions.

The studio technology they had at their disposal was initially very limited, but they were able to extract the most out of what was available by working with people such as engineer Tom Dowd, who later became a major producer himself.

In the early fifties, recording studios were generally using mono tape machines. The possibilities for manipulation were limited: different takes could be spliced together, and it was possible to add tracks only by re-recording what was already on tape along with the new material, which meant losing a generation every time a track was added. For this reason, Leiber and Stoller as well as other producers rehearsed their acts well before going into the studio, because what was performed in front of the microphones was what the record was going to be.

The duo had such success with their novel ways of "writing" records from conception through final mix that they became executive producers for the recordings of others. What this meant, according to Stoller, was that they taught Brill Building songwriters how to make a record, i.e., how to work in the recording studio. Things got quite technically specific; there was a lot to know, and details often made a significant difference in the final sound of a track. For fine tempo adjustments, for example, there was the technique of taking a recording and "speeding it a wrap," meaning putting tin foil around the spindle of the take-up reel on a tape deck, thereby causing the tape to pass over the heads faster and causing a slight but perceptible increase in speed of the recorded material. Such seat-of-the-pants techniques were later incorporated as standard, precisely controllable features on professional tape machines. It was and is quite common for producers to suggest new features to equipment manufacturers, who are in a constant race against each other to come up with the most up-to-the-minute product.

Yet even the pioneers had their moments of doubt about some of the most radical new developments in recording technology. When the first eight-track recorders arrived in 1958, Jerry Leiber is said to have exclaimed: "Oh no! We're going to lose our souls!" There was indeed much apprehension about losing the live "feel" so important to the rock sound, but after an initial period of resistance the new technology was embraced by the entire industry.

Leiber and Stoller not only became producers (or "supervisors," as they were first termed on LPs by the Drifters), but also were instrumental in getting the producer's role acknowledged and credited, both on the record sleeve and at the bank:

Stoller: We fought for that. We fought with our good friends at Atlantic [ . . . ] who at first said "But you've got your names on the record already, you're the writers" [ . . . ].

Leiber: But we were changing, and the business was changing, and we thought [producing] would be a legitimate involvement. It's a legitimate job we're doing, so why shouldn't a credit be fixed to this job, just like the director of a film, or the cutter of a film? [ . . . ] Whoever's doing whatever has some kind of function, so why don't we get a credit for this? [ . . . ] We knew we were performing the job of an A&R man at one of the big companies who was pulling down a very big salary to perform this function. We were doing it, but we were not receiving more than the songwriter's royalty, so we were doing two jobs [ . . . ]. I think we were the first to break the ice, creating a situation where a credit was given and a royalty established. (Ibid., 18)

## Phil Spector's Wall of Sound

Into this brave new world arrived a nineteen-year-old *wunderkind* from Los Angeles named Phil Spector, who was to inscribe the term "producer" firmly and indelibly into the vocabulary of pop record-making, and who would be remembered by the chroniclers of rock'n'roll as one of its most talented, eccentric, and flamboyant figures.

Leiber and Stoller had taken him on as an apprentice as a favor to their friend Lester Sills, a mover and shaker in the L.A. pop music scene. Spector had already had a hit on the West Coast with "To Know Him Is to Love Him," but the real action was in New York. Like many before and after him, Spector did not enjoy instant acceptance in the tightly knit, highly competitive world of New York studio musicians. He had to prove himself, which he soon did with "Corrine, Corrina," surprising those around him with his command of the studio and his knack for arranging and selling his sounds. The song, sung by Ray Peterson, reached the Top Ten in late 1960.

Spector was soon one of the hottest figures in American pop music. He cultivated a Superman image, boosted by press from the likes of a young and enthusiastic Tom Wolfe in his 1963 essay "The First Tycoon of Teen":

He does the whole thing. Spector writes the words and the music, scouts and signs up the talent. He takes them out to a recording studio and runs the session himself. He puts them through hours and days to get the two or three minutes he wants. Two or three minutes out of the whole struggle. He handles the control dials like an electronic maestro, tuning various instruments or sounds up, down, out, every which way, using things like two pianos, a harpsichord and three guitars on one record; then re-recording the whole thing with esoteric dubbing and over-dubbing effects—reinforcing instruments or voices—coming out with what is known throughout the industry as "the Spector sound." (Wolfe, 43)

Not all of this is true. Like most producers, Spector had a group of regulars he worked with. Engineer Larry Levine and arranger Jack Nitzsche are consistent presences on Spector's productions from 1962 to 1966, his period of greatest success, and his writing credits are mostly shared with established Brill Building songwriting teams such as Gerry Goffin and Carole King or Ellie Greenwich and Jeff Barry.

Spector's first exposure to what may have inspired his famous Wall of Sound was in working for Leiber and Stoller, who pretty much ignored their charge and went about the business of making hits. During the period that Spector was working in their studio and sleeping on their couch, Leiber and Stoller were already working with multiple instrumentalists in their rhythm sections in order to achieve a fuller sound:

We started building this rhythm section. By the time we were in full swing we were using like three to five guitars—a twelve-string, a lead guitar, two rhythm guitars [ . . . ], then we used up to three percussionists and a drummer. (Quoted in Fox, 173)

This technique of using multiple instruments playing the same role in the studio was later to evolve into Phil Spector's trademark Wall of Sound. Leiber and Stoller are careful not to take credit for Spector's achievement, and draw a distinction between their approach to the overall sound of a recording and his:

Phil was the first one to use multiple drum kits, three pianos and so on. We went for much more clarity in terms of instrumental colors, and he deliberately blended everything into a kind of mulch. He definitely had a different point of view. (Quoted in R. Palmer, 28)

There was indeed quite a difference in production styles. Recordings by the vocal group the Drifters from the period 1958–1962, which were produced by Leiber and Stoller, sound more conventional than Spector's records of roughly the same period. Even as they move away from doo-wop and toward an early soul feel, the sound of the Drifters records is grounded in jazz and even classical idioms. And although their 1959 hit "There Goes My Baby" is credited with being the first R&B record to include strings, the sound of the orchestra is not as striking as Spector's. The production on these Drifters recordings is more about the notes and less about setting a mood or painting a picture with timbre. The beginnings of a massed sound are there, but the instruments are still equal with the vocals; voices and instruments share the same space, and trade roles from time to time over the course of the song. The focus is also primarily on the ability of the players, on well-played solos and nifty twists and turns in the arrangement, not on the unusual overall sound-picture which is the hallmark of Spector's Wall of Sound.

So what *is* this Wall of Sound that dominated the sound of American pop music between 1960 and the arrival of the British invasion in 1964? In the most obvious sense, it is simply putting a lot of instrumentalists in the recording studio and having them all play at once. Consider, for example, the recording personnel for "River Deep, Mountain High," recorded with Ike and Tina Turner in 1966 (Ribovsky, 221):

four guitars

four basses

three keyboards

two percussionists

two drummers

two obbligato vocalists

six horns

a full string section

In addition, according to keyboardist Billy Preston, "[Spector] had every machine going all at once and it was a circus and he was the ringleader" (ibid., 155). Another source lists similar forces as the norm for the Spector productions of the early sixties:

The instrumental backing tracks on classic Phil Spector productions were normally provided by three drummers, three bass players, numerous guitarists and keyboard players, a three- or four-piece horn section and several percussionists. (Tobler and Grundy, 50)

A more complete definition of the Wall of Sound would add to the obligatory massed instrumental and vocal forces the fact that it incorporates an augmented R&B rhythm section, i.e., rhythm guitar, drums, and bass playing particular patterns; as well as the liberal use of echo chamber and tape echo effects, derived from Sam Phillips's Sun Records "Memphis Sound" (ibid., 52).

Overall, the technique was a celebration of sonic grandeur, achieved by both physical and technological means. By taking total artistic control of a recording, Spector in fact redefined what it meant to produce a record. In addition—and perhaps most important—Spector had a vision for what the total sound or effect of the recording should be, and he knew how to achieve that effect by manipulating the available technology. He changed forever the way the producer's role would be viewed.

### "Be My Baby"

One of the best examples of Spector's sound is the 1963 hit "Be My Baby," sung by the Ronettes and featuring Spector's then-wife Ronnie Spector as lead vocalist. At 2:37, this track epitomizes the concept of the concise pop song: there is a honed simplicity, an inevitability about each element of harmony and structure that is emblematic of the genre. The arrangement as well as the mix exemplify the Wall of Sound at its finest. Recorded at L.A.'s Gold Star Studios, where Spector had recorded his first hit in 1958 before moving to New York to work for Leiber and Stoller, "Be My Baby" was engineered by Larry Levine and arranged by Spector's regular collaborator Jack Nitzsche, with writing credit going to Spector, Greenwich, and Barry.

The track opens with two bars of a sparse but powerful R&B drum beat awash in cavernous ambience. On closer listening it turns out that the massive effect of the snare, which occurs only on the four until the chorus, is achieved not only by a hard hit and lots of room tone, but also by the addition of percussion (woodblock and tambourine) as well as hand claps. On the one of the third bar, the rhythm section comes in full force with the addition of emphatic shakers on every eighth-note, as well as a bass section and a piano section playing a three-note figure in counterpoint to the beat. Persistent octave Es, again on every eighth-note, augment the feeling of forward motion. In addition, prominently placed castanets begin describing little airborne trajectories that begin somewhere after the three of every bar and land exactly on the snare's massive four.

The lead vocal begins in the fifth measure, and is striking for a number of reasons. First, there is the unaffected yet seductive performance of an appealing young female voice singing of romantic passion; from a production standpoint, the placement and imaging of the voice in the mix are masterful. The effect of the primary room tone, which is created by placing the microphones at a distance of somewhere between ten and fifty feet from the performing musicians, is to place the instruments at a distance from the voice. There is also some ambient coloration of the vocal, but it has a more subtle quality, and its shorter duration (faster decay) and lower proportion to the primary source signal (i.e., the voice) place it closer to the listener on the virtual stage. The overall effect is intoxicating, as though a special someone were hypnotizing you with her mere presence and you were disconnected from reality. The words are sung against the backdrop of a sonic landscape suggestive of fantasy and desire as the song builds toward the climactic chorus. A horn section dominated by a baritone sax enters and underscores the tension of the dominant by outlining a descending sequence which leads to the return of the tonic E.

When the chorus arrives, it is a powerful release. Not only are we back at the tonic, but a galvanizing backbeat now appears: the snare finally fills in the two of every bar, now really propelling the arrangement. A choir comes in, taking over the melody the voice sang during the verse, thus freeing Ronnie to embellish the main theme of the song. The castanets also change, going into a busy flamenco-style pattern, which is placed considerably further back in the mix.

Fifty-six seconds into the track, it becomes clear that the chorus was just a stage in what is to be a terraced buildup which occurs throughout the song. We come back to the verse and its lower energy level, although by now most listeners will crave a return to the power of the chorus, which is still suggested in form of the choir. The background voices continue to underline the lead vocal with chordal "ooohs" and "aaahs," keeping the energy level up and the listener in a state of expectation. When the chorus returns, it repeats its earlier statement and is then followed by a "middle eight" in which a slightly bizarre string section takes over the lead line— one hears what sounds like a blend of violas and cellos in the high register. After another eight bars of regular chorus, the song comes to an abrupt halt as the drums return to their opening statement for two bars. Then, as suddenly as they took over the spotlight, they are again joined by the wash of instruments and voices of the chorus. As the song fades, the drums play inventive breaks every other bar.

One of the best-known features of Spector's approach to recording was his obsessiveness. He is known to have made the Crystals rehearse for hours, even weeks at a time in the dark singing "Uptown," pushing them to the point of physical exhaustion. Similarly, Tina Turner tells of having been pushed to her limit while recording "River Deep, Mountain High"; and the eight hours Spector spent in an L.A. studio making Johnny Ramone play one guitar chord over and over for the opening "Rock'n'Roll High School" on the Ramones' *End of the Century* are the stuff of rock legend (Bessman, 111).

But there is more to Spector's insistence on hearing something time and again in the studio than mere eccentricity. He was using the studio as his orchestra, arranging the timbres of various voices, instruments, effects, and room tones in much the same way as a more conventional composer would employ the colors of the orchestra. This is where his genius lies, and it is what inspired a great many producers who followed him. It is what makes Spector the quintessential pop-rock producer: for him, the point is not just the melody but the overall *sound*, the feel of a recording.

Every composer knows that it is difficult to predict exactly what the sound of a given arrangement will be, especially if one is attempting something out of the ordinary; and every engineer knows that the exact placement of the microphone, the room tone, and the settings of the mixing board have a huge effect on the sound of a track. Today many pop composers and arrangers rely on some form of a MIDI mock-up in an effort to preview the actual sound of their work, as is evident from artist interviews in trade magazines such as *Keyboard*, *Mix*, and *Musician*. In the early sixties, the only way to know what all the ingredients of a record were going to sound like when put together was to put everybody in the studio and try it out. Few people were fortunate to have such means at the time; Spector was one of them. To him, the studio was a musical instrument, to be tuned and practiced on and performed with. The degree to which he took this idea was considered excessive by some at the time. Yet listening to his recordings, it becomes apparent that what sets him apart from his predecessors and contemporaries is the innovative, unique timbre and feel of his productions.

It is easy to imagine that a percussion arrangement such as the one created for "Be My Baby" could be a cluttered mess if all of the elements were not balanced exactly right. The placement of the castanets described earlier is one example: they are featured prominently in the verses, and are practically hyperactive in the choruses. What makes all this activity acceptable in the context of the whole is the exact placement in the mix. The miking and ambience of just these little percussion instruments is nothing short of brilliant—they glide effortlessly from bar to bar as if propelled by a magic power. This attention to detail, to multiple levels of instrumentation and color and placement in the mix, is the sort of thing Spector excelled at in the studio. And he seems to have alienated just about everyone by trying just about everything:

"You just didn't know what was gonna happen until he had his guitars playing," [recording engineer] Levine said. "They'd play the first four bars over and over and he'd have 'em play fifth notes and then change it to sixteenths [sic]. He'd start the guitars, add instruments, then take 'em away and start the guitars all over again. Everyone else was waiting for his turn, but the guitars never got a break. [Guitarist] Howard Robbers played for hours on end that day, and I remember him saying that his wrist was killing him." (Ribovsky, 121)

Yet the importance of Phil Spector to the development of the function and influence of the producer is immeasurable. The most visible element of his influence was the cult of personality which grew up around him. He was the first "star," brand-name producer. Perhaps more significantly, he was the first producer to develop a strikingly original sound which the public associated him with: the Wall of Sound.

Spector can be thought of as the link between the shift away from the "realistic" recording aesthetic of the fifties to the innovations of the sixties. The move was away from a method in which the sound of a recording was created primarily by the instrumental arrangement and its live performance in the studio, toward one where the manipulation of technology is at the very least an equal partner in achieving the final sound of the recorded work. Whereas the craft of the studio technician and producer had formerly been to create for the home listener a perfected version of an artist, band, or orchestra in concert, the rules of the game were now changing: the object was no longer to create a flawless "real-life" experience, but rather to use the available technological resources imaginatively in order to create sounds that were no longer functioning within the metaphor of realism which had previously been the norm. Thus, by the mid-sixties, "manipulating technology" had come to encompass a whole lot more than mic placement or fader levels.

## Brian Wilson

Another brilliant producer who came to prominence in the sixties was Brian Wilson of the Beach Boys. A founding member of the band, Wilson refused to tour after 1964, choosing instead to devote himself full-time to writing and recording the band's music. The Beach Boys, formed in 1962, had become one of the most successful acts in the world with light songs about California beach-bum life ("Surfin' USA," "Help Me Rhonda," "I Get Around"). Initially, their long-playing records followed the standard practice of the time, which was to put out an album with ten or twelve songs which didn't necessarily have anything to do with each other. But the ambitious, elaborate *Pet Sounds* (1966) broke with that tradition, and is today considered by many to be the first-ever concept album. It put Wilson on the map as a major innovator in the field of music production. He has said that he felt challenged to outdo the Beatles' *Rubber Soul*, and Paul McCartney and George Martin have both acknowledged the influence of *Pet Sounds* on the making of *Sergeant Pepper* (Leaf, 109). Wilson, in turn, had a nervous breakdown around the time *Sergeant Pepper* appeared in 1967, which was apparently due to a combination of excessive LSD consumption and the realization that he had been outdone by the Beatles juggernaut (ibid., 56).

A great admirer of Spector, Wilson made it his business to emulate his production techniques, to the point of working with Spector's hand-picked group of session musicians, the Wrecking Crew, in the same studio (L.A.'s now-defunct Gold Star Studios), and with the same engineer (Larry Levine). Just to make sure he had the right idea, Wilson also took some lessons in record production from Spector. The following passage is taken from a 1996 interview with Brian Wilson:

Brian: If you want to know the truth, I think I was trying to emulate Phil Spector in some ways with my tracks.

Interviewer: Let's talk about that. There is a clarity of instrumentation on your records vs. the "Wall of Sound." Can you tell me what it was about Spector's work that impressed you?

Brian: Well, he had it in his mind. He knew in his head what he wanted before he got to the studio, obviously. What had happened was, I'd been called down to Gold Star a couple times by Phil Spector, and I think he really wanted to teach me a little bit about production. I didn't know that at the time because I was just a young, naive little guy. But later on I realized he was there to help me, and he was there to teach me about something.

Interviewer: Of what you watched, learned, and observed, how did it specifically influence you as a producer?

Brian: I learned that he would say "Let's hear the electric piano," and he'd walk toward the window in the booth and he'd look out and he'd go, "That's what I want." "That's good." "All right, let's do this." Like in Creation. "It is good." What it really came down to is he taught me how to create records [ . . . ].

Interviewer: You once talked about how you learned from him that if you had a piano and a guitar and combined them together that created—

Brian: A third sound. You get a different sound. I used that in *Pet Sounds*. "I Know There's An Answer"—you'll hear an organ and a tack piano together. It's really neither of those sounds. It's like a completely new sound. A different sound. Combining one thing to make another thing. It's amazing. (Ibid., 10)

Although the Beach Boys' label, Capitol Records, was uneasy with the new direction embodied by *Pet Sounds,* they did release it; thirty years later, the label furthered the understanding of the pop production process by releasing a wonderful three-CD set called *The Pet Sounds Sessions.* This remarkable collection provides highlights from the instrumental tracking sessions, as well as the vocal tracks alone, which they call "Stack o' Vocals" (disc 3). In addition, the album is provided in two forms: the original mono, and a more recent stereo mix.

The care with which *Pet Sounds* was made is legendary. Many of the participants have reminisced about Wilson's perfectionism; Mike Love, one of the Beach Boys, coined the nickname "Dog Ears" for Wilson, because "he was hearing things none of us could hear" (ibid., 16). Today, it is clear that the perfection of the vocals is in a class by itself, and that the vocal tracks alone make for an enjoyable and engrossing listening experience. Writer Kingsley Abbott notes that

The discipline and the complexity of the vocals on *Pet Sounds* is easily recognizable. Unlike the Beatles and the Rolling Stones, who bent vocals and took liberties without censure, the Beach Boys remained extremely structured, developing Brian's incredibly sophisticated vocal parts. (Abbott, 73)

The fact that the vocal tracks of *Pet Sounds* are in a class by themselves can easily be confirmed by listening to the "Stack o' Vocals" disc mentioned earlier. The complete blend of voices, the utter precision of intonation, and the exquisite ensemble of attack, phrasing, and expression are everywhere evident. Perhaps the most amazing selection of all is the a cappella "God Only Knows," which in its intricate and beautiful vocal work evokes the image of Brian Wilson as a sort of Perotin of pop. One cannot help but wonder just how such extraordinary craftsmanship and aesthetic beauty was achieved. Derek Taylor, publicist for the Beach Boys at the time, wrote this account:

After swallowing the lumps in their throats and exchanging uneasy glances, the work begins. First Al and Bruce step up to the mike to do the *dumdedums*, then Carl, Al, and Dennis for the *runrunweeoos*. And Carl and Bruce for hums with Brian on falsetto. And on and on into the night, overdubbing, rearranging, softening, strengthening, shifting voices, moving Al further away, Dennis back a step, Carl closer, Mike lower. Patiences wear out. Brian will accept nothing less than perfection. (Quoted in Abbott, 73)

"The harmonies that we are able to produce give us a uniqueness," Brian explained, "which is really the only important thing you can put into records—some quality that no-one else has got. I love peaks in a song—and enhancing them on the control panel. Most of all, I love the human voice for its own sake . . ." (ibid.).

While studio technology was moving quickly forward, both of Brian's favorite studios, Western and Gold Star, had only four-track recording machines:

With the notable exception of his use of Columbia's new eight-track technology for "California Girls," Brian would record all the instrument tracks onto one of the four tracks and use the remaining three for the vocals. During the *Pet Sounds* sessions, with the help of engineers Chuck Britz and Larry Levine, Brian would sometimes use two four-track machines, subsequently transferring the sounds onto an eight-track machine. This would allow Brian seven tracks for overdubs. (Ibid., 72)

*The Pet Sounds Sessions* collection also includes two instrumental cuts from the tracking sessions for the hit single "Good Vibrations," another classic which was recorded as part of the *Pet Sounds* sessions but not released on the album itself. The making of "Good Vibrations" is a case study in perfectionism:

The first takes of "Good Vibrations" were recorded on 18 February, 1966 at Gold Star Studios. . . . No fewer than sixteen more sessions were to follow at Gold Star, Sunset Sound, Western and Columbia between 9 April and 1 September, at a reported cost of between $50,000 and $75,000, before Wilson [was satisfied]. (Cunningham, 79)

Working in the studio with Brian Wilson must have required exceptional patience. Mike Love writes of recording upwards of 30 takes in just one section of backgrounds for "Wouldn't It Be Nice." [ . . . ] If one gives "Pet Sounds" or "Good Vibrations" a listen, we can safely say that the vocal performances are as close to perfect as humanly possible. (Quoted in Leaf, 81)

One of the most interesting things about Brian Wilson is that he was both the songwriter and the producer of *Pet Sounds*. In this sense, he was building on the work of Leiber and Stoller. Yet Wilson really wanted to realize the full potential of the recording studio. He was not only interested in directing performances, but was also involved in every detail of the production of the sounds themselves. Wrecking Crew electric bassist Carol Kaye remembers that "Brian would take over the console once [engineer] Chuck Britz had done a preliminary setup, then he directed us from the booth via intercom or waving motions" (ibid.). In another interview, Kaye recalls that

Brian always brought written out charts for most of the musicians. [ . . . ] He wrote the charts himself, you could tell from the illegibility of them sometimes. The notes were sometimes on the wrong side of the stems etc. He didn't hire a professional copyist like the rest of the arrangers did. (Quoted in Abbott, 57)

Already in 1966, then, the composer, arranger, and producer are melded into one person, prefiguring a practice which has become increasingly common in the contemporary recording studio. Brian Wilson was at the controls himself, making on-the-spot decisions about notes, articulation, timbre, and so on. He was effectively composing at the mixing board and using the studio as a musical instrument.

## "Good Vibrations"

"Good Vibrations" is a clear example of the use of the studio as a compositional device. The fact that it is so clearly contrived, that its seams show so obviously, does not detract from its appeal as a song. Listening to "Good Vibrations" today, one can immediately hear that it is spliced together from recordings made in different spaces. Yet the music is powerful, a heady mix of exceptionally polished vocal harmonies and experimental instrumentation.

The song opens with the verse, sung by the coauthor of the song, Mike Love. His flawlessly double-tracked vocal is accompanied by Carol Kay's bright, plucked Fender bass, Hammond organ, flute, drums, and percussion. The unusual register and timbre of the electric bass make it sound more like a sort of baritone guitar, while the choice of stops and use of echo chamber give the organ an unusual quality. Sixteen bars (or eight slow bars) into the song, at 0:25, there's a very obvious splice in the first chorus. The ambience, indeed the whole quality of the sound changes, but it doesn't hurt the flow; the change is so intriguing that one is led to suspend disbelief. The resulting effect is much like cutting away from one shot to another in film.

The drums are transformed completely, going in an instant from a full drum kit with bongos and large ambience to minimal, dry tom-toms. The tambourine stays, but its jingle shifts to a higher register and its ambience changes, indicating a different instrument being played in a different room. The organ and the flutes disappear and are replaced by a theremin playing a countertheme to the vocals, as well as a cello playing ostinato F triplets. Additionally, a shift to double time combines with the change of the lead vocal from falsetto to baritone to create a sense of forward motion. This makes for an effective contrast between the dreamlike, calm quality of the verse and the very animated chorus.

There are also simple yet effective structural symmetries at work in "Good Vibrations." The verse and chorus, so different at first glance, both rely on simple sequences. In the verse, the bass, organ, and flute all play a simple, four-bar sequence which descends by steps (E♭–D♭–C♭–B♭). The movement is reversed in the chorus: here, the vocals carry a sequence which rises by whole steps (G♭–A♭–B♭). But while the harmonic rhythm is regular in the verse, the rate of change in the chorus is divided into the proportion 2:1:1. Either way, we end up on the dominant, and the song stays there for a while after the second chorus, paradoxically creating a dreamy sort of tension (1:41). The unusual combination of harpsichord and honky-tonk piano gently insist on B♭ in a sort of arrested blues progression while the vocals tell us that "I don't know where / but she sends me there."

When the bridge is followed not by a return to previous material but by another bridge, this time in F (the dominant of the dominant), the magical quality of the song intensifies, and the ear may begin to wonder just where this music is going to end up: we expect a return, but are given an unexpected twist instead. The dreamy, tense quality is heightened, and a satisfying, full bass and harmonica take over as the vocals fade out. It all seems to indicate "lights out" until the sudden appearance of the revelatory seventh chord heralding the return of the upbeat mood of the chorus, which we had almost forgotten about. After a rollicking bit of stacked vocal harmonies, which feature an airy falsetto episode, the song returns to the theme of the chorus minus the vocals and quickly fades out.

As we have seen, the production of "Good Vibrations" sharpens the outlines of its structure, and also provides variety and interest to a simple framework. The changes of orchestration and acoustic space from section to section serve to take the listener into new sonic realms every thirty seconds or so, which may be one reason Wilson dubbed this song his "pocket symphony": the listener is presented with quite a variety of material in just over three and a half minutes, and the song has a consistent quality of forward movement and surprising, felicitous change. Unusual combinations of instruments (theremin and cello, harpsichord and honky-tonk upright piano) and an almost fanatical attention to relative levels (the various mix positions of the cello triplets in different parts of the song, for example, or the subliminal theremin in the first bridge) are combined with imaginative shifts of timbre and exceptional vocal performances for a lively, unusual, compelling track, more than thirty-five years after its creation. Although Brian Wilson never quite recovered musically from his nervous breakdown following the completion of *Pet Sounds,* his place in pop history is assured by the startling innovations and fresh, appealing sound of the recordings he made in Los Angeles in 1966.

## Hits Off the Assembly Line: Motown

The pop music factory known as Motown records was designed and built by Berry Gordy beginning in 1959. For a little over a decade, his Hitsville Studio at 2648 West Grand Boulevard in Detroit churned out hit after hit, giving the world songs by artists such as Stevie Wonder, the Jackson 5, the Supremes, Smokey Robinson and the Miracles, the Temptations, and Marvin Gaye. Producers included the team of Holland–Dozier–Holland, Norman Whitfield, and Smokey Robinson.

From the beginning, Gordy cultivated an assembly-line approach to making music, a concept he famously culled from his own experience as a worker in a Ford automotive plant. His idea was to run a recording empire not unlike the movie empires of Hollywood studios. In the studio, singers worked with a coterie of musicians known as the Funk Brothers, a group of changing membership akin to Phil Spector's Wrecking Crew. Different producers would sometimes get assigned to work with the same artist; which track was to be released would then be determined at long meetings with producers and company executives present. Gordy was at the helm of the whole operation, overseeing the creation and promotion of a style, a big-picture approach to making popular music, more than individual songs:

He was supremely confident that the basic sound he had created was the key to success, with the singers almost being secondary. When asked if his label was producing a sound and not songs, he told a Detroit reporter: "You probably haven't any voice. But there are probably three notes that you can sing. I can take those three notes and give them an arrangement and some lyrics. That makes a song. And your song will sell." (Posner, 52)

Berry Gordy was an entrepreneurial producer who seems to have been as comfortable behind his desk as at the mixing board. Yet he was also known for spending long hours in the studio:

Gordy was usually at the mixing board, blending and remixing. "Mixing was so important to me that it seemed I spent half my life at the board." Smokey Robinson and others teased him that he was a "mix maniac," but Gordy countered that while the differences in many mixes were subtle, they could make or break a record. (Ibid.)

When one thinks of the Motown sound, the Detroit era recordings come to mind. This was before the company moved to Los Angeles, and its distinctive sonic signature became absorbed into the world of L.A. pop-studio perfectionism.

While some writers (notably Jon Fitzgerald) have written extensively about the rhythmic and tonal patterns of Motown songs, there is also a distinctive sonic character to Detroit Motown recordings that sets them apart. It's a particular quality of the midrange, and of harmonic distortion in the upper portion of the frequency spectrum: individual sounds often have a particular brightness, even a brittleness associated with them. There's almost always a sizzle to the Motown sound, a kind of subtle distortion that would be considered unacceptable in contemporary recording practice; it somehow projects performance energy, and it's also something that speaks well through small speakers. Writer Gerald Posner notes that

From his Ford Motors work, [Gordy] was fascinated with a way people were increasingly listening to popular music: on car radios. He experimented mixing songs so they sounded great on those tinny players. Mike McClian, Motown's chief engineer, built a small radio that approximated the sound of those in cars. Everyone listened to early mixes on that radio. . . . [the music] sounded good on small transistor radios, the popularity of which was expanding quickly in the early 1960s. (Ibid., 51)

Physical properties, then, or the materiality of the medium, are an influence on sculpting the frequency space, and thus on what can be called the electronic orchestration of a recording at the mixing board. This "knob-twisting" aspect of production is often overlooked, especially in the literature. But as Gordy points out, it can be crucial to the sound of a track. It is also important to keep the notion of *sound* in mind when attempting to define the Motown (or any other production's) characteristics. In the case of Motown, the very room they recorded in, and the way the equipment was calibrated and used, played an important role in the overall effect of the music. Gordy is fully cognizant of this. In a 1990 *Rolling Stone* cover story interview, he relates how once in 1959 he was unhappy with the sound of his studio, so he took the Miracles to Chicago to record in a state-of-the-art recording facility. What he found was that the new, "improved" recording, which he liked very much, failed to sell, whereas the original version had quickly sold 60,000 copies. His experience taught him that "the first version had a certain honesty about it. It wasn't slick. So after that, we continued to produce songs in our own little studio. . . . [it had] real wonderful acoustics and a magical sound, but we just never fully recognized that at the time" (quoted in Goldberg and Seeff).

To grasp this vital aspect of production is to understand an important facet of the role of the producer in the studio. He is not only listening to the notes and whether or not a singer is giving his best performance; the producer is also listening for the total effect of the music, and must have an understanding, if only intuitive, of all the elements that go into making a track. Room acoustics are highly complex, as are the many distortions and realignments undergone by the electrical signal that a microphone sends out to a tape machine before it reaches the listener's ear from a loudspeaker. There is no such thing as a truly "natural" sound; there are only subjective opinions. This is why engineers, producers, and audiences (although the latter are generally unaware of specifics) favor certain rooms and equipment over others: while they cannot necessarily name just what it is that makes a certain room sound good, hearing is believing. In Berry Gordy's words, "we weren't concerned about whether it was right or wrong, we just wanted to know if it sounded good" (Gordy, 3).

In recent years, the role of the Funk Brothers in creating the Motown sound has been explored in some detail. Perhaps the most visible of these recent reappraisals is the film *Standing in the Shadows of Motown*, which features interviews with, and performances by, the surviving members of the Funk Brothers, the informal name of the group of Motown session players. The film is valuable in that it brings the excellent musicianship of the people behind the star image to the fore, and gives the players long-overdue credit for their contribution to the magic of Detroit-era Motown. But the role and contribution of the producer is left unexplored. Throughout the film, the producers remain mysterious, somewhat adversarial mystery figures behind the double glass. Separated from the producers acoustically by necessity, the musicians apparently did not hear what discussions went on in the control room; they were only given direction once the discussions between the producer, engineer, and others in the control room had taken place.

At the heart of the matter lies the issue of who should receive credit for the Motown catalog. The argument is often made that studio work is a collaborative art. This is true to a great extent, but it is important not to lose sight of the big picture, of the overall vision necessary to bring about a context for people to work in, both in terms of physical and conceptual space. Viewed in this light, it is clear that Berry Gordy is the person who created Motown and guided it, and that he and his staff producers were the ones ultimately responsible for the sounds that were chosen and formed into Motown recordings. But it must be kept in mind that the thesis that the producer has gradually become a composer, utilizing the studio as a musical instrument, is not contradicted by acknowledging the contributions of engineers, musicians, arrangers, and songwriters. Gordy himself says that if he had come along at a later time and had become a rapper, "I would have been doing all my own songs because I wouldn't have needed the voice quality. [ . . . ] there wouldn't have been any need for me to have artists" (quoted in Goldberg, 67).

### "I Heard It through the Grapevine"
Produced by Motown regular Norm Whitfield, this 1968 recording sounds thoroughly modern, yet also somehow dark and mysterious. As one writer puts it, "Grapevine" is "an extraordinarily sophisticated recording that nevertheless seems to go back a good four hundred years for its dark,

utterly ominous incantations" (Anthony and Henke, 273). Blues keyboardist Mike Bloomfield simply refers to the song as "voodoo music"
(ibid.). The song stands out as one of the best pop recordings ever made,
despite having been recontextualized and overexposed by its use in a variety of commercials for everything from raisins to cars.

To engage this song by close listening is to discover a plethora of fasci-
nating details, some no doubt accidental, some planned. As the recording
date recedes into history, what probably seemed like straightforward
recording practice at the time takes on special meaning as the technology
and recording conventions and practices of the period are rendered obsolete by historical development.

One can almost see Hitsville's musicians squeezed into the tight confines
of Studio A, also known as the Snake Pit, at 2648 West Grand Boulevard,
Detroit. There is a liveness, an organic cohesion to the recording that creates a unique mood. This happens right from the opening double hit of a
tambourine and a snare that sounds like it is being tapped with brushes.
The distinctive electric piano riff in the foreground is complemented by a
more atmospheric Hammond organ holding down a quarter-note pulse.
The lightly tapped bass drum has an unusual amount of tone, and ushers in
every bar, playing a composite beat together with the hi-hat. The rattle of
taut snares is heard in sympathetic vibration with both keyboards, indicating live performance of multiple instruments in the same physical space.

The song builds by adding electric guitar and a high electric piano
voice doubling the first riff a tenth higher. The shake of the tambourine is
reminiscent of a rattlesnake; it sounds ominous, threatening, and evokes
Old Testament imagery (the snake, temptation, sin). The tension is
heightened to a first breaking point by a dramatic French horn octave
glissando. This dovetails with the entrance of the vocal and James Jamer-
son's characteristic Fender Precision '62 bass (0:23), and the first verse
begins.

The beat switches to a tight interlock of syncopated yet laid-back,
behind-the-beat playing on toms and congas. The figure is complex and
sounds like the drummer is playing the loose tom skins lightly, in perfect
concert with the congas. This percussion section sounds like one large percussion instrument; the interlock brings to mind West African drumming.

The opening riff that built so swiftly has been transformed into a more
laid-back presence in the form of the bass. Continuity is built into the

arrangement by having the opening figure continue, and there is also variation provided by having it switch to the bass, with its characteristic lower
frequency range and monophonic voice. The vocals carry forward the
opening instrumental complaint (which is also an invitation to sublimate
by dancing), adding a new dimension to the mood by voicing feelings of
anguish, jealousy, and fear of the loss of the beloved. The string section
provides supple ornamentation, a touch that gives the track a polished surface that contrasts with the more complex, at times rough timbre of the
vocals.

At the pre-chorus (0:39), the female background vocalists appear, echoing Gaye's sentiments and providing a link between the strings and his
voice as the tension builds again, this time heralding the chorus.

Unlike other songs of the period ("Be My Baby," for example, or "Good
Vibrations"), the arrival of the chorus at 0:47 is not so much a dramatic
event as simply the logical consequence of all that has gone before it. The
opening electric piano figure and its subsequent restatement by the bass
turn out to have been foretelling the full stop in the middle of the word
"grapevine," thus tying the elements of the music together in yet another
way. Taken together, the many subliminal congruities in the song are what
gives the listener the sense that it coheres utterly.

The second verse (1:11) reiterates the first, followed by another chorus;
at 2:04, a brief instrumental turnaround is carried by the strings, leading
into the third and final verse. After a final chorus, the music comes to an
end as the backing voices intone the ostinato fourths heard briefly twice
before in the song. They do indeed sound like a voodoo incantation, and
the sense of dark, ominous magic is heightened by Gaye's free vocalizing as
the sound fades away.

Whitfield's production of this track is clearly a vital part of its character.
The sonic congruences I've discussed are fixed by virtue of their having
been recorded and mixed, rather than notated on paper. Thus the recording replaces the written score as the definitive artifact; the difference is not
unlike that between a finished film and a movie script.

### George Martin and the Beatles

Another towering figure of pop production in the sixties was George Martin, producer of the Beatles. Working out of London's Abbey Road Studios,
he collaborated with John, Paul, George, and Ringo to produce some of the

greatest recordings of the century. Martin began his career in the early fifties at EMI recording classical music, but he soon moved on to making comedy records. As the head of Parlophone, a subsidiary of EMI, he acquired experience as a producer working with comedian–actors Peter Sellers, Dudley Moore, and Peter Ustinov, among others. At the same time, he began to branch out into pop recording, and eventually he had a number of hits on the English charts with singers such as Cliff Richards and Cilia Black in the early sixties.

Martin first got together with the Beatles in 1962, recording their first single, "Love Me Do," which didn't make it past number seventeen in the English charts. Their second single fared better. "Please Please Me," released in January of 1963, quickly reached number one, and in Martin's words: "from that moment, we simply never stood still" (Martin and Hornsby, 130).

Martin's evolving role as the Beatles' producer closely resembles the long-term development of the function of the producer in the industry as a whole. In the early days of working with the Beatles, he writes,

my function as a producer was not what it is today. After all, I was a mixture of many things. I was an executive running a record label. I was organizing the artists and the repertoire. And on top of that, I actually supervised the recording sessions, looking after what both the engineer and the artist were doing. Certainly I would manipulate the record to the way I wanted it, but there was no arrangement in the sense of orchestration. They were four musicians—three guitarists and a drummer— and my role was to make sure that they made a concise, commercial statement. I would make sure that the song ran for approximately two and a half minutes, that it was in the right key for their voices, and that it was tidy, with the right proportion and form.

At the beginning, my specialty was the introductions and endings, and any instrumental passages in the middle. I might say, for instance: "Please Please Me" only lasts a minute and ten seconds, so you'll have to do two choruses, and in the second chorus we'll have to do such-and-such. (Ibid., 132)

All of that was soon to change. With the huge commercial success of the Beatles, they and Martin were free to do as they pleased; awed by their success, no record executive was going to stand in their way. True to the era they lived in, the Beatles (especially John) took psychoactive drugs, LSD in particular. In an era of general permissiveness and encouragement of all sorts of experimentation and alternative everything, even the commercial marketplace was a viable forum for presenting challenging sounds reflecting the tumultuous spirit of the decade. And although he eschewed controlled substances, Martin already had a reputation for experimentation of the sonic kind with his comedy records. His interest in electronic music and strange sounds in general was more than passing:

Creating atmosphere and sound pictures . . . that was my bag. I did a lot of it before the Beatles even came along. In 1962 Parlophone issued a single called "Time Beat/Waltz in Orbit," a compilation of electronic sounds, composed by a certain "Ray Cathode"—me!

Just along the road from us at Abbey Road, the BBC had set up an experimental sound department, called the Radiophonic Workshop. I was fascinated, and got to know some of the engineers down there: people who spent their entire time cooking up freaky sounds, with whatever they could lay their hands on. They had the (to them) standard equipment, of oscillators and variable speed tape machines, but they also indulged in quite a bit of concrete music. (Martin and Pearson, 83)

Before discussing the groundbreaking musical experiments of the period from 1966 to 1968, it is necessary to take a look at the state of recording technology at the time. At the beginning of the sixties, most EMI product was recorded on two channels. Two-channel machines were available, but they were not generally used to record stereo; most releases were still in mono. Martin recorded all of the instrumental tracks onto one mono track, and then all the vocals onto the other track. It then became possible to mix the performances offline. In an era of increasingly hard and loud sounds, of electric guitars and drums replacing pianos and strings, this mixability of individual channels was a most desirable development. It meant that one did not have to worry about the voices being drowned out by blaring instruments. It even meant that one could record the vocals separately from the instruments, and punch in on their track only.

In late 1963, EMI installed one-inch, four-track recorders at Abbey Road Studios. Although this was a huge improvement, it was still quite limited. When Martin asked for more tracks, the European manufacturers told him that one-inch tape could not accommodate more than four. Curiously, Martin seems to have been unaware of America's eight-track Ampex recorders, which existed as early as 1958. The way Martin's production team (principal engineer Geoff Emerick and seconds Richard Lush and Phil McDonald) got around the limitation of having only four tracks to work with was to use multiple machines. It must be remembered that in the days before digital recording, there was a loss of quality each time a track was copied. The inevitable result was added tape hiss and distortion. Initially, Martin limited himself to simply mixing down four tracks onto a two-track stereo lathe. There was a standard production procedure:

The first [track] took the drums and the bass. To that you would add, on a second track, the harmonies, which might be played by guitars, piano or something else. The lead voice would go on a third. The fourth track would be for the little extra bits—what today we call the "sweetening." (Martin and Hornsby, 149)

It is often said that *Sergeant Pepper* was "made on a four-track." The phrase has become something of a rebuke aimed at those who engage in overproduction or excessive complaining about the limitations of their equipment ("we only had sixteen tracks," etc.). Here's what George Martin has to say about the four-track techniques employed in the making of *Pepper*:

One technique was to mix four of your original tracks down to what would normally be the finished two-track stereo product. You would then put those two tracks onto a fresh four-track machine, which would leave you two spare "open" tracks to play with. [ . . . ] If you were very courageous, and you wanted a very heavy sound on one particular track, you could dub all four original tracks down to that one. [ . . . ] Now you had three other tracks with which to sweeten it. When it later came to the making of the *Sergeant Pepper* album, that technique was taken almost to absurdity. (Ibid., 150)

There were up to three four-track machines involved in the making of *Sergeant Pepper*. It was not until 1967, after that album was finished, that Abbey Road took delivery of its first eight-track machine.

As we have seen, the early recording career of the Beatles followed a conventional path. The record which marked the turning point in their use of the recording studio was *Revolver* (1966). From this album through the *White Album*, Martin and the Beatles experimented with abandon. They dispensed with the concept of realism or what could be called "figurative" recording, often constructing instead a virtual or imaginary space unconfined by what is possible in the "real" world of live performance on conventional instruments. For Martin and the Beatles, the performability of *Revolver* and the groundbreaking records that followed it was not a concern, not only because the Beatles were not obliged to perform live (they retired from touring in 1966), but because the records themselves succeeded in shifting the audience's expectations from a replicated concert experience to something more internal.

### "Tomorrow Never Knows"

The first track to break definitively with the past was "Tomorrow Never Knows," the last song on *Revolver*. While the album as a whole contains numerous unusual production ideas, notably the string arrangement on "Eleanor Rigby" and the sound effects on "Yellow Submarine," "Tomorrow Never Knows" was a total departure from anything the Beatles had attempted before.

The song consists of three main elements: the hypnotic, riveting osti-nato of Ringo Starr's drums, coupled with the bass, unchanging through-out the entire song; a well-selected assortment of tape loops fed to the faders of a mixing console; and John Lennon's vocal.

An opening sitar drone fades up at the beginning of the song. It is soon joined by drums and bass playing an obstinate pattern. While the drums play only the subtlest of variations in shading, the bass changes its groove slightly over the course of the song, switching octaves in the middle of the instrumental break. The bass is finger-picked, and its soft, rounded tone contrasts with the forward-moving octave-C groove it holds down throughout.

The drum sound has especially subtle timbral features. The snare, which is only heard on the "two," has a low-level reverb tail, which gives the lis-tener the feeling of being right next to something located in a very large, reverberant space. A crash cymbal enters with the snare and tom and is played in an unusual manner: rather than being "ridden" to fill out the beat or struck as an accent, the cymbal is tapped continuously on its edge. The technique creates such a wash that the stick attack is not heard; rather, it is a pleasing sort of pink-noise drone which blends with the sitar and is heard throughout. A prominently placed tambourine answers the snare on every "and four" at the beginning of the track, somehow managing not to conflict with the tom-toms, which fall on "ah-four."

Eight seconds into the song, a mysterious sound that vaguely evokes seagulls with a Doppler effect is heard for the first time. This is the first of a number of tape loops which are featured prominently in the song and give it its unique character. As George Martin explains, "We had all these loops coming through the mixing board continually, so we were able to just raise any fader and hear the sound of that loop. It was a kind of electronic organ" (ibid., 164). This technique represents an important step in the use of the recording studio as a musical instrument. It prefigures the advent of digital sampling and looping by about twenty years. Writer Ian Macdonald gives a detailed account of the loops employed:

There were five in all, each running on an auxiliary deck fed onto the multitrack through the Studio 2 desk and mixed live: (1) a "seagull"/"Red Indian" effect (actually McCartney laughing) made, like most of the other loops, by superimposition and acceleration (0:07); (2) an orchestral chord of B flat major (0:19); (3) a Mellotron played on its flute setting (0:22); (4) another Mellotron oscillating in [6E] from B flat to C on its string setting (0:38); and (5) a rising scalar phrase on a sitar, recorded with heavy saturation and acceleration (0:56). (Macdonald, 152)

Loops are playing loops, as the Mellotron was itself a loop-playback device invented in the early sixties. A piano-style keyboard triggered multiple loops when played polyphonically, thus making low-budget string sections and other simulations possible. These were in turn recorded onto tape for "Tomorrow Never Knows" and altered (sped up, slowed down, or reversed). They were then made into continuous loops as was standard at the time: a piece of tape set to the proper tension was run around one or more microphone stands rather than from reel to reel, with the capstan and pinch roller moving the tape past the playback head.

The vocal enters on the fifth bar, outlining a second-inversion C-major chord. It is the voice of John Lennon, exhorting the listener to "Turn off your mind / relax and float downstream." It has a slightly surreal quality, which was created by another technique invented at Abbey Road called "ADT" (automatic double tracking). This effect was achieved by an early use of a tape recorder's record head being used for playback as well as recording. When played back in combination with the signal from the playback head, the gap between the two heads created a delay which could be time-varied according to the speed at which the tape was played back. This is the same principle used by the once widely popular effects unit known as the Echoplex.

On the fifth bar of this and almost every verse, an organ enters on B flat and is joined by an orchestra loop. Each time there is a slight variation; the orchestra enters first in one instance, another time there is no organ. At the end of the song, the B flat finally resolves to C every four bars as the music fades out.

The instrumental break offers more new sounds. Its phrasing is unusual in that the customary eight or sixteen is divided four–two–four–four–two rather than the usual four–four. Also noteworthy is that the material for the break consists of a combination of tape loops and Paul's solo from "Taxman," "slowed down a tone, cut up, and played backwards" (Martin and Hornsby, 152). Martin's team and the Beatles succeeded admirably in making the whole a credible, even compelling musical experience. The "treated sitar" loop which makes the first statement of the instrumental break clearly forms a rising antecedent phrase, which is in turn answered by a two-bar consequent fashioned from the "Mellotron flute" loop (both of these are so altered as to bear little or no resemblance to the acoustic instruments which were originally recorded). The inimitable but often intractable sound of backwards electric guitar is also presented in a well-phrased, listenable form, which leads out into Lennon's voice again, this time run through a pair of rotating Leslie speakers which were miked in the studio.

For all its *musique concrète* virtuosity, what is most striking about the song is that is indeed a song. There is a conventional underpinning consisting of intro–verse–break–verse–fade; it has a steady, danceable beat; the vocal is melodic and easily intelligible. Although the piece is devoid of choruses and is harmonically static, it still can be considered a pop song, with all the airplay and mainstream success working in the genre can entail. In this, it differs from "Revolution 9" on the *White Album*, which does not adhere to any pop conventions. One of the reasons many people consider *Revolver* as a whole to be the finest Beatles album is that it succeeds so well at integrating a wide variety of styles and influences, as outlined above.

### "A Day in the Life"

Integrating disparate sound sources into a coherent whole has increasingly become one of the producer's most important tasks. But perhaps no one has faced greater challenges or succeeded more completely in this regard than George Martin. Not only did he and his production team have far fewer technological resources with which to accomplish their work than are standard today; they also had to contend with factors peculiar to the Beatles. In at least one case, Martin had to help devise a way to combine two separate songs into one ("A Day in the Life"). On another, he was called on to blend two vocal performances sung in different keys and at different speeds into one ("Strawberry Fields").

"A Day in the Life" is intriguing for a number of reasons. According to George Martin, this was the first song the Beatles worked on which would become part of *Sergeant Pepper*. Like "Tomorrow Never Knows," this song ended up as the last track of the album it was released on.

The song opens with a simple acoustic guitar in the left channel. It is soon expanded upon by straightforward piano chords and an electric bass which pan back and forth in a brief crescendo, leading into Lennon's vocal, which is placed on the right channel. Entering at the same time as the voice, a pair of close-miked maracas provide an eighth-note ostinato on the left channel. Then, in what is surely one of the most subtle drum arrangement/mixes in all of pop, a distant, muted, barely audible drum provides subliminal accents emanating from the left channel at the end of the first two phrases. The big entrance of the drums comes after the line "He blew his mind out in a car," providing both a dreamlike, once-removed sound picture of the crash and propelling the arrangement toward the break.

The drums also have a sound which has never been heard before or since. The snare and cymbals sound real enough, but the tom-tom sound falls somewhere between fat and timpani-like. This was probably achieved by slowing down the tape for the recording of the middle and lower tom-toms; it is hardly possible to actually come up with such a drum sound off a real drum set at 1:1 speed. Aside from the sound, the character of the drum part is also highly unusual. Rather than simply keeping time, the drums are assigned a role which falls somewhere between the function of percussion in the classical orchestra, which is primarily one of punctuation, and how drums are usually heard in a pop song, holding down a steady backbeat.

The two ascending orchestral glissandi which occur in this song (pair of songs?) are legendary:

Paul had been listening to a lot of avant-garde music by the likes of John Cage, Stockhausen, and Luciano Berio. He had told John he would like to include an instrumental passage with this avant-garde feel. He had the idea to create a spiraling ascent of sound, suggesting we start all instruments on their lowest note and climbing to the highest in their own time. (Martin and Pearson, 56)

Martin wrote out an orchestration of Paul's concept, hired forty-one orchestral players, and stacked five different performances of his arrangement on top of each other to create the effect of a huge orchestra. In addition,

Geoff was tweaking away at the studio's built-in "ambiophony" system, which instantly fed the music back through the hundred speakers spaced around the studio walls to create a customized—and highly amplified—sound. (Ibid., 59)

Knowledge of these unusual techniques and equipment helps explain why the orchestral glissando, which makes its first appearance one minute and fifty-three seconds into the song, still sounds so extraordinary thirty-three years after it was recorded. Aside from its sheer sonic power, the gliss acts as an aural vortex which transports the listener from the dreamlike mood of the first part of the song into the workaday reality of the double-time section sung by Paul. This was originally a fragment of a separate song, which was integrated into "A Day in the Life" after the outer verses had already been recorded. Apparently John had simply not known what to put there and asked Paul for help. George Martin was called on to make it all work together technically and arrangement-wise.

An alarm clock heralds Paul's "hurry-up" vocal, which is also imaged to the right channel. The acoustic guitar is now absent; the piano has taken over the comping, while the shakers, bass, and drums heard in the first verse return. Paul's unmistakable bass is mixed rather high as usual, while Ringo's drumming is restricted to a more standard time-keeping function. Soon John's voice reappears singing a wordless melody in G; the orchestra enters and holds pedal tones beneath him, leading the section convincingly to a close and segueing directly into a reprise of the first verse. All of the elements of the first part of the song reappear, and again the listener is led into a huge orchestral ascent which is very similar to the first but this time features a more active drum part by Ringo. The rich sonic texture culminates in a huge piano chord, which was "played by Lennon, McCartney, Starr, [road manager Mal] Evans, and Martin on three pianos, and multitracked four times" (MacDonald, 183). The fadeout is rather long at thirty-seven seconds, and was obtained by compressing (electronically "squeezing together" or limiting the signal strength of) the original chord and then gradually bringing up the faders as the signal slowly faded away.

As with "Tomorrow Never Knows," one of the main achievements of "A Day in the Life" is the successful integration of avant-garde elements into the form of the pop song. Both productions share a similar, rather simple structure, which is essentially *ABAB*. The surprises are the *"B"* sections of each song, consisting of the unique sound fields described above. In each case, the Beatles and their production team managed to do the impossible, which is to make plausible the juxtaposition of such disparate elements. This joining of opposites is evident not only in the songs mentioned above but on the two albums as a whole. In the case of *Revolver*, for example, there are such clearly delineated yet disparate stylistic influences as soul

("Got to Get You into My Life"), chamber music ("Eleanor Rigby"), East Indian ("In My Life"), pop ballad ("Here, There and Everywhere"), and of course *musique concrète* ("Tomorrow Never Knows").

Such stylistic diversity is rarely successful, and it is unique to the Beatles, especially in comparison with other albums of the time. One thing that makes *Sergeant Pepper* stand out is that there is again such a wide range of musical influences at the same time even as they are presenting one of the first full-blown, unified concept albums, replete with psychedelic-retro costumes. There are other pretenders to the distinction of having been the first concept record, notably Phil Spector's 1963 *Christmas Album,* Frank Zappa's 1965 *Freak Out,* and the Beach Boys' 1966 *Pet Sounds,* but the Beatles went further than their competitors.

George Martin's imaginative production techniques were an important part of the success of the album. Once the Beatles had decided on an alter ego (Sergeant Pepper's band), it fell to Martin and his team to actually stage the kaleidoscopic sonic reality the album presents. The many disparate sonic elements had to be tied together with a lot of production skill. Martin's work influenced countless rock producers; competitors were listening closely, especially in California.

## Frank Zappa

Frank Zappa was obsessed with the activity of recording as much as he was with composing. Like Brian Wilson, he was primarily interested in recording his own work. He acted as producer for practically all of his prodigious compositional output from 1960 to his untimely death in 1993, releasing more than seventy CDs. He worked with some of the best musicians and engineers of his time; his compositional output ranged from writing for his own ensembles to working with the Ensemble Intercontemporain and the London Symphony Orchestra. Zappa also spent three years in the mid-eighties composing exclusively on the Synclavier, the legendary early digital synth-sequencer (Michie #1, 4).

Zappa, who was based in Los Angeles, often employed two engineers in shifts in his $3.5 million home studio. He was an avid, hands-on editor. "Zappa could edit like nobody could," writes his longtime engineer Mark Pinske. Working on remixing the L.S.O. recordings, "I think we had about 1,000 edits. . . . I was privileged to have learned from somebody like that" (quoted in Michie #2, 2). When he became too exhausted to hear accurately, he would disappear into his anechoic chamber for a few hours to catch some sleep, then reemerge to resume mixing. Zappa once was asked whether he considered the master tape or the score the definitive document; he answered "the tape." This stands to reason, as much of his output combines instrumental and real-world sounds with studio processing and editing to create the overall texture. This is especially true of early works such as *We're Only in It for the Money* (1967), which was recorded in New York at Apostolic and Mayfair Studios. The album's cover is a parody of *Sergeant Pepper*, while the lyrics skewer hippies and authoritarianism with equally sharp barbs.

The album usurps the tasteful Beatles stylistic mélange of *Revolver* and *Sergeant Pepper*. It's a compelling mixture of stand-up comedy, elaborate *musique concrète* experimentation in the manner of Pierre Henry, surf music, social satire, throwaway sixties pop, and avant-garde orchestral textures. In terms of the approach to recording and production, the work is full of innovations. It is a uniquely postmodern pop album, in that it repeatedly breaks the frame within which music is usually presented. For example, engineer Gary Kellgren's voice suddenly appears at various times, whispering paranoid rants. The whisper, heavily treated with multiple-echo delay, is the engineer's running interior monologue about the making of the record, as well as his fantasy about how "one of these days I'm gonna erase every master in the world, leaving nothing but blank, empty space." (The device of recurring commentary is a precursor of the documentary-style voices heard musing about the moon on Pink Floyd's *Dark Side of the Moon*.) By making the listener aware, however satirically, of the processes involved in making records, the album becomes a vehicle for illuminating what goes into its making, its materiality.

One of the qualities that most distinguishes Zappa from other pop producers of his time is his extraordinary ability as a composer of *musique concrète*. Famously inspired by Edgar Varèse, Frank Zappa built meticulous, well-thought-out tape pieces. He interspersed them, at times even commingled them, with his satirical songs, and often used tape collage to build segues from song to song, a device the world first heard on 1965's *Freak Out*.

The opening track of *We're Only in It for the Money*, "Are You Hung Up," immediately makes clear that the Mothers of Invention, Zappa's name for his group at the time, are about more than casually experimenting with avant-garde ideas. The first twenty-five seconds of this prologue to an extraordinary concept album present a sonic collage second to none. It is a studio construct par excellence, in that it instantly sets up a number of layers of voices and bizarre, unidentifiable noises, and then sustains the whole bizarre arrangement in a convincing manner.

Taken as a whole, the tableau is quite elegant, with subtle timbral relationships among the various noises, as well as a particular rhythm and a quality of forward motion, even though there are no identifiable musical instruments present other than the human voice. Low growls and almost ultrasonic high pitches frame a recurring, variously delayed vocal sample that can best be described as someone hemming and hawing; its midrange has been boosted, making it sound like it's coming through a radio or megaphone. This is transformed into the voice of Eric Clapton asking: "Listen, are you hung up?" a few times, each varied in a kaleidoscopic—psychedelic—way. A chorus of munchkins descends into the middle of the frequency space, and the focus shifts to the voice of the aforementioned paranoid-delusional sound engineer. Just as suddenly, bad surf music is heard, to the sound of breaking waves—but then the frame is again broken, this time by the painful sound of a record-player needle being dragged across a record. Zappa once told David Marsden that "we took a Davey Jones record that the engineer happened to have sitting around, and we put it on a Gerrard turntable. We spent approximately half an hour trying to get the right kind of record scratch. Then when we got the one that sounded good we panned it back and forth" (quoted in Courrier, 140).

Zappa's sensibility was experimental, yet he was working in the context of pop music. He once described himself as "filling in the holes between pop and classical music." Zappa's sonic canvas is really the totality of sound, both "musical" and "nonmusical." If John Hammond was essentially a facilitator, and Les Paul a tinkerer, Zappa was a hands-on eclectic, crossing cultural boundaries and integrating diverse musical styles and sensibilities in his sound workshop. His early work differs markedly from that of contemporaries such as the Beatles and the Beach Boys in that it is ironically self-aware and categorically embraces all sound. The work uses abstract sound not only as an enhancement or embellishment of a song, but as an end in itself.

Zappa did not limit himself to abstract sound effects and sonic collage. He also raised the bar for rock performance standards. In 1969, he collaborated with the exceptionally gifted keyboard and woodwind player Ian

Underwood. The result was the seminal *Hot Rats*, which paved the way for much ambitious jazz-rock of the early seventies. Zappa's compositions incorporated a combination of live, "natural" acoustic instruments with sped-up, slowed-down, and processed sounds, now mostly tonal in origin.

For Zappa, the tape recorder was also something of a time machine, a locus for transforming reality. One of his favorite techniques was something he called "xenochronous" recording, a technique which involved taking individual tracks from different live recordings of his group and matching them up to create unlikely mergings of harmonies and especially rhythms. This was possible thanks to the Mothers' recording truck, purchased from the Beach Boys. The truck and the stage setup were standardized so that over a period of years, the EQing and levels of the instruments matched perfectly from concert to concert. In addition, Zappa always made sure that his guitar solos were recorded separately (either on a separate Uher recorder or line-in), so that they could be recycled onto new material written in the studio.

### "Flower Punk"

The dark, avenging passion of Jimi Hendrix's "Hey Joe" has been transformed into a parody of hippiedom in San Francisco, circa 1967 (Hendrix's recording is itself a cover of a song by Billy Roberts, recorded in 1965 by L.A. garage band the Leaves).

The first sound heard on the track, which is the tenth cut of *We're Only in It for the Money*, is a classic Zappa segue: the mellow, floating 6/8 texture at the end of the previous track, "Absolutely Free," is cut off with a brief but rude burst of noise that sounds like a wrench being tossed into a wheel of a passing subway train. This event ties in with the numerous sound collages heard throughout the album, and simultaneously forms the opening downbeat of "Flower Punk."

The opening texture is one of drums, electric bass, and what sounds like an electric keyboard run though a wah-wah pedal. About four seconds in, two unnaturally boyish voices (one in each channel), possibly the sped-up voice of Frank Zappa, enter with a repeated question: "Hey punk, where you goin' with that flower in your hand?" The song has been rebarred into alternating groups of four bars each of 7/8 and 5/8. The order and ratio of riff and chord progression are inverted. The original chord progression of "Hey Joe," C–G–D–A–E, is condensed into the four quick bars of 5/8 at the end of each phrase; the A–D–E riff heard at the end of each phrase of "Hey Joe" is pressed into service as the main figure of "Flower Punk." Hendrix drummer Mitch Mitchell's freely flowing, midtempo drumming becomes in the Zappa version a manic, virtuosic alternation between kick and snare, accompanied by a hi-hat that transfers to a splashy ride cymbal (or ride-as-crash) as the song progresses. Billy Mundi's tight, stripped-down, four-square sound foreshadows later drumming such as hardcore metal (speed and accuracy) and jungle (fast, snare-kick-cymbal only); its focused energy and precision shine through the poor recording quality.

"Flower Punk" is obsessively repetitive; for about the first minute and a half, the listener is pummeled with the two quickly alternating figures described above. But at 1:33, things suddenly take a turn for the bizarre. In the lyrics included on the album cover, one reads that "just at this moment, the 2700 microgram dose of STP ingested by FLOWER PUNK shortly before the song began TAKES EFFECT." Zappa is mocking psychedelia, while at the same time creating an elaborate, and deranged, psychedelic tableau. The voices, speaking simultaneously in hard-panned separate channels, now begin to spin complementary fantasies of rock stardom, and how it will bring fame, money, and the attention of the opposite sex. The bass pounds away on a pedal C (a reference to "Tomorrow Never Knows"?) over driving drums, while munchkin voices and strange intermittent bursts of pitched noise on each channel round out the sonic landscape.

In the late eighties, Zappa reissued his early work. Owing in part to the deterioration of master tapes, there was an extensive reworking of the original masters. On *We're Only in It for the Money*, he even went so far as to re-record the bass and drums. These tracks, recorded by engineer Mark Pinske, both have a very mid-eighties sound that does not blend well with the idiomatic, charmingly lo-fi, mid-sixties sound of the original elements of the mix. Pinske reports that "we would spend two or three days [ . . . ] to try to get the absolute best drum sound. [ . . . ] when we were done, we would have a really great, elaborate-sounding drum set. So Frank started really liking this really good drum sound, and [ . . . ] wanted to start hearing it on just about everything. It was just a phase" (quoted in Michie #2, 5). As Steve Lindeman has noted, "there is an incongruity of timbres in this amalgamation" (Lindeman, 2). One thing that is clear is that even though it is the engineer who gets the actual sounds, what ends up on the released recording is ultimately the producer's responsibility. This re-recording, along with reworkings of *Hot Rats* and other albums, also raises issues about which recording is the definitive one: the first version released? Or is it the artist's later re-recordings and remixes? Once the artist (and the producer) releases the recording, does it belong to him or to society at large? Zappa's answer was that his music is in a continual state of becoming, that there is no "definitive" version. This seems reasonable, since he was a master editor, given to reworking and reassembling his creations throughout his life.

Frank Zappa is important to the development of production because he was a true sound sculptor who knew how to create soundscapes, instrumental compositions, and skillful blends of the two. Gifted with unusually discerning ears, he was present at every stage of the creation of his sonic artifacts. Some began on paper, others as fragments of tape, still others as improvisations on stage that were later spliced together and even layered between performances. An obsessive denizen of the recording studio, Zappa's inquisitive nature and sonic imagination raised the bar for rock, jazz, and classical recording alike.

### The Situation at the End of the Sixties

The rise of the long-playing record and the development of multitracking coincided with the increased visibility of the producer in the recording industry as a whole. Figures such as Leiber and Stoller paved the way for the stardom of a Phil Spector primarily as a producer of hit singles. It was not until the wide proliferation LPs and the equipment to play them on that the need for a producer became acute. LPs, which meant more time to fill and the possibility of larger forms and multitracking, which brought with it both the promise and challenge of offline production, made it desirable to have someone in charge of managing the new expanse of fifteen- to twenty-minute sides and multiple tracks.

As usual, pop music grew in tandem with the technologies available to it. Where the first pop LPs were compilations of (often previously released) singles, after a couple of years the idea of relating songs to a central theme became popular. Thus 1966 saw the release of the Mothers of Invention's *Freak Out* and of the Beach Boys' *Pet Sounds*; *Sergeant Pepper* was released in 1967, and was followed by many others, notably the Who's *Tommy* and

David Bowie's *The Rise and Fall of Ziggy Stardust*. Each of these albums has the name of a widely known producer associated with it, whether it be Frank Zappa, Brian Wilson, or George Martin. And in each of these albums, there is a blurring of the distinction between producer, arranger, and composer.

There were also other forces at work aside from the technological and artistic. The mid-sixties were of course a time of tremendous upheaval and change. It was also a time of innovation and invention, a time when new ideas in art were seemingly everywhere. Pop and rock music in particular were to youth culture what computers are today—that is to say, matters of central importance. Just as the youth of today embraces the speed of inno vation in computing which makes ever-faster games and cooler Web sites feasible, so the counterculture of the sixties could not get enough new and more experimental sounds which reflected and indeed played an important part in developing their collective outlook. There was a confluence of expanding audiences eager for truly new sounds with significant advances in technology. Coupled with strong economic expansion in the Western world, this resulted in ideal conditions for musical exploration and innovation. As technology became more central in people's lives, pop music followed suit. Listeners embraced the bold experiments of the Beatles and their colleagues, thus encouraging further growth.

The sixties were the beginning of the use of the studio as a true musical instrument. By the end of the decade the importance of the producer was something every professional pop musician and even a great many fans were aware of. On both sides of the Atlantic, a professional cadre was developing, and a slew of recently invented studio techniques were in the process of becoming part of the standard repertoire of professional recording. With the rise of complex technology, the pop producer became an indispensable interpreter between the technical and the artistic aspects of making records. His ability to shape sounds, even create entire performances, had already had a profound impact on popular music.

## 2   The Studio as Musical Instrument

### Sixteen Tracks and More

The artistic scope and inventive studio technique of concept albums such as *Sergeant Pepper*, the Beach Boys' *Pet Sounds*, and the Mothers of Invention's *We're Only in It for the Money* modeled an expanded palette of sonic expression for the popular music world of the seventies to build on. The primacy of the 7" single had been eclipsed by the 12" LP, with its maximum duration of over twenty minutes per side, and music fans eagerly awaited new releases by million-selling acts such as Led Zeppelin, the Who, Jimi Hendrix Experience, Pink Floyd, and a host of others.

Sonic experimentation was in. Tape manipulation on at least one of the tracks of a hit album by a major guitar-based rock band was common (Led Zeppelin's "Whole Lotta Love," Hendrix's *Axis: Bold as Love*), and the synthesizer was now a featured instrument on songs by major rock bands (the Who's "Baba O'Riley," Pink Floyd's "One of These Days").

Producers, engineers, and the acts they worked with began to use the expanded capabilities of the recording studio to extend it into a device for arranging and composing. Major recording studios reinvested some of their profits from previous projects into a phalanx of new recording equipment. Sixteen-track recording, one of the first usages of which was by Frank Zappa in 1969 at L.A.'s Wally Heider Studios, became widespread in professional studios by 1971, and twenty-four-track recording was introduced in 1973. This radical and rather sudden expansion of possibilities completely changed modern recording techniques and sounds, as recordings of the period show. As Mark Cunningham points out:

The additional flexibility given to engineers through the use of sixteen-track machines explains the sudden clarity of rhythm tracks, particularly drums, which came to fruition in the early Seventies—a period when strict close miking was the order of the day and little or no room ambience would creep into the mix. (Cunningham, 176)

The increase in tracks available to people working in the professional recording studio brought with it both increased sonic fidelity and expanded creative possibilities. The most creative acts and producers enhanced both aspects of the evolving art of recording, weaving conventional songwriting and instrumentation together with sound effects, imaginative use of the latest studio outboard gear, and deft splicing and elaborate mixing to create vibrant new records.

### Dark Side of the Moon

One of the most successful productions of the early seventies was Pink Floyd's *Dark Side of the Moon*. A phenomenally successful record by any standard, this recording remained on the charts for twenty years, selling millions of copies since its release in 1973. It is that rarest of productions, a commercially successful album which is at once highly listenable, technically and conceptually innovative, and artistically engaging.

From its inception, Pink Floyd was an experimental band. Just about everyone who was anyone, including the Beatles, made a pilgrimage to London's Notting Hill Church during 1967 to see one of Pink Floyd's multimedia performance pieces. At the time, the driving conceptual force behind the group was Syd Barrett, who was soon to become one of the casualties of drug abuse as a result of an overindulgence in LSD. Under his behind-the-scenes direction, Pink Floyd made some very inventive and influential records, such as *Piper at the Gates of Dawn* (1967).

Recorded only five years after *Sergeant Pepper* and in the same studio (London's Abbey Road), *Dark Side of the Moon* adopts the idea of the concept album. Yet it is a very different production. The most important technical advance is the use of two sixteen-track machines, whereas *Pepper* had used three four-tracks (see chapter 1, p. 29). The band, which is credited on the album as having self-produced, and engineers Alan Parsons (tracking) and Chris Thomas (mix) pointed the way for excellence in recording. Their work remains sonically first-rate today, almost all of its production decisions having stood the test of time. It is also notable for its integration of live and electronic elements, as well as its use of ambience as a thematic element.

By the time of *Dark Side*, Barrett had been replaced by guitarist David Gilmour, who led the group in an uneasy alliance with bassist Roger Waters (their enmity finally led to a split in 1979). Yet it can hardly have been just these two men who made the album sound the way it does. The indirect influence of the Beatles is everywhere evident: not only did Pink Floyd record this album at Abbey Road Studios, they also drew on people who had worked with the Beatles themselves. Alan Parsons had engineered *Abbey Road*, and Chris Thomas was an uncredited producer and musician on the *White Album*. Moreover, the producer on *Piper at the Gates of Dawn* had been Norman Smith, who worked with the Beatles as an engineer before becoming a staff producer at EMI. *Piper* had been recorded at Abbey Road at the same time as the Beatles were finishing up *Sergeant Pepper*.

The function of the producer had undergone a radical transformation in the sixties, and by the seventies it was clear to most people in the music business that production had become a major part of any recording project. Thus, besides Pink Floyd, artists such as James Brown and Jimi Hendrix credited themselves with production, and artists began to produce other artists: Paul McCartney produced folk singer Mary Hopkin, and David Bowie produced Iggy Pop's *Lust for Life*. The term "producer" came to mean anything from wide-ranging technical expertise and arranging skills (George Martin) to simply being a figurehead with a big name (Andy Warhol "producing" the Velvet Underground). Ever since Leiber–Stoller and George Martin had successfully sought producer royalties, a credit as producer on a hit record meant money in the bank, and a few producers seem to have claimed credit for financial gain.

Yet while increased royalties may have been the motivation for some, the members of Pink Floyd could credibly claim advanced production skills by the time of *Dark Side*. Guitarist Gilmour was one of the first to use his own effects setup, rather than relying on the studio engineer to shape his sound and come up with novel textures. The group had only gradually weaned itself from Norman Smith, crediting him as producer on their first album (*Piper at the Gates of Dawn*, 1967) and executive producer on *Atom Heart Mother* (1970), with themselves listed as producers. They finally began taking full production credit as of *Meddle* (1971). The band had also had extensive exposure to electronic synthesis though its collaboration with electronic composer Brian Geesin on *Atom Heart Mother*. Pink Floyd seem to have been like sponges, soaking up techniques and ideas from those with whom they worked, eventually becoming totally self-sufficient once they found out what they needed to know. To their credit, they transformed what they learned from others into their own distinctive style, and utilized the recording studio to help shape their unique sound.

In discussing *Dark Side of the Moon*, it seems appropriate to use a somewhat lower level of magnification than that applied to the musical discussions in the previous chapter of this essay. As any skywatcher can attest, when one switches to lower magnification on a telescope, what one loses in detail is to be balanced against the larger picture. The album itself seems to demand this approach owing to its metabolism, which is one of relatively simple tableaux welded together with great skill.

Pink Floyd were practicing for *Dark Side of the Moon* years before it was made. Thus the chiming clocks heard on track three refer back to the (decidedly lower-fi) clocks heard at the end of *Piper at the Gates*, while the sonic grandeur of some passages recalls *Atom Heart Mother* (1970) and *Ummagumma* (1971). Generally, the ideas and procedures of the band's previous albums have been refined and stylized into the tight production that is *Dark Side*. As of *Meddle* (1971), the pompous horn and string charts of *Atom Heart Mother* are out, replaced instead by shimmering guitars, tasteful keyboard and synthesizer work, and well-thought-out sound effect textures. The very slow tempos, simple progressions, and tasteful slide guitar work are already present on *Meddle*, awaiting the elaboration and refinement heard on *Dark Side of the Moon*. There is still some straight-ahead blues on the former album, and it doesn't have the balanced proportions of *Dark Side*. *Meddle* is made up of five separate songs which clock in at six minutes or under, while the last track (originally side two of the album) lasts for 23:31. Conceptually as well as sonically, *Meddle* simply lacks the coherence of *Dark Side*. The producers at times still seem to be coming to grips with how to record, placing for example the kick and snare on opposite channels on track three of the record. By contrast, *Dark Side of the Moon* is a mature product, and as such is not weighed down by the uneven quality of the early stages of experimentation.

The album opens with a slow fade-in of a heartbeat (Nick Mason's gated kick drum), fading into a sonic tableau, "Speak to Me/Breathe," which includes speaking, cackling, and screaming voices, a loop of a ringing cash register, and a modulated synthesizer sound. Each of these elements is a

sonic motif which will reappear at least once later in the course of the continuous, forty-two-minute work. When the curtain opens and the band
begins to play (1:18), they do so with what even today is pristine sonic clarity, every detail of guitars, percussion, and bass totally present and beauti-
fully distributed across the stereo field. Contributing to the impression of
balance and harmony is the unique "floating" feel laid down by drummer
Nick Mason and bassist Roger Waters, a rhythmic underpinning which is
enhanced by the graceful, multitracked slide work of David Gilmour. There
are also multiple high-fidelity reverbs at work, foregrounding the vocal
against the instruments by placing it in a somewhat smaller space. At
exactly four minutes (beginning of track two, "On the Run"), the music
hard-splices to an abstract Hammond–Leslie texture, which is soon joined
by multiple synthesizers to create an anxiety-inducing, resonant-filtered
sonic tableau which is distinctly cinematic—a reflection of the band's
many film soundtracks recorded over the course of the previous four years
for Michelangelo Antonioni and others.

As the ominous clouds of filtered noise recede, a plethora of clocks
appear, chiming chaotically; the heartbeat returns as well. After a virtuosic
Rototom solo by Mason over long, held notes by Gilmour, the shift of
focus back to the band playing a song featuring Roger Waters's vocals is
accompanied by a startling change in ambience. This paradoxically serves
to smooth the jarring effect of going from an abstract texture awash in
ambience to a very "daytime," rational consciousness. The technique
recalls the shift on the Beatles' "Day in the Life" from the surreal orchestral
ascent (with an alarm clock signaling the change) to the workaday feel of
the "woke up / fell out of bed" section discussed in chapter 1: the listener is
moved from a very large space to a tight, constricted one, the change in
ambience being used as an arrangement device to indicate a change of
scene and mood.

There's also a difference between the music beginning at track two ("On
the Run"), which suggests the specific image of someone running through
dark city streets while pursued by a helicopter, and the music beginning at
track three (0:55), which does not evoke such specific associations, as it
uses neither sound effects nor musical instruments to suggest real-world
sounds. *Dark Side* moves here nimbly from cinematic tableau to sound
effects to psychedelic instrumental to song. The underlying assumption,

inherited from *Sergeant Pepper*, is that we are not in a real theater at all; the reality of the illusion is seamless and nearly complete.

The next change of scene in this audio movie is into a piano progression, punctuated by one of a number of barely intelligible vocal interjections which recur throughout the album and are accompanied by Gilmour's ethereal slide guitar. These spoken fragments were apparently culled from answers to the question "What is on the dark side of the moon?", recorded by Waters and flown into the tracks at various points. The spoken word fragment is followed by a long crescendo with a wordless vocal by soul singer Clare Torry.

"Money," the fifth cut of the album, is the only track separated by silence from the previous material. This is because the album was originally released as an LP, which was by nature limited to about twenty minutes per side at full bandwidth (it was possible to press sides up to an absolute limit of about thirty minutes at 33 1/3 rpm, but the overall volume and bass response were reduced). The physical limitations of vinyl as well as the CD have influenced the pacing of records. Most obviously, a pop single's "B" side tended to contain more experimental, noncommercial material, while an LP would be paced to include an intermission about twenty minutes in. Since the advent of the CD, sides have vanished, and the demands on recording artists have grown: forty-five minutes is now considered the minimum acceptable duration for a CD (LPs were OK at thirty), with total playing times of around an hour or more considered preferable.

The use of ambience as an arrangement device continues on "Money." Its tight, close-miked, all-dead-surfaces sound is a strong contrast to the soaring music at the end of side one. Only the vocal is given a tight, slap-back ambience. When the sax solo enters, the ambience is increased slightly; the guitar solo (3:06) moves suddenly from the audio equivalent of the aspect ratio of television to widescreen cinema, but the music shifts gears again at 5:10. The vocal returns, dissolving into a more elaborate collage of spoken voices. These in turn lead in to the next track, "Us and Them." Here the mesmerizing opening feel of the band playing as if suspended in midair returns.

As in most rock music, the pulse of the music has subtly quickened as the set progresses, even in this lower tempo range, where quarter notes go by at anywhere from 66 to 75 beats per minute (bpm). *Dark Side of the*

*Moon*'s pacing is extraordinarily deliberate; shifts of 3 or 4 bpm tend to happen between phrases, not during them. Moreover, the overall tempo is unusually low for a rock album. *Sergeant Pepper*'s first track opens at 95 bpm, with the reprise clocking in at a healthy 125; *Revolver*'s ballad "Here, There and Everywhere" was recorded at 82, while Sly and the Family Stone's early seventies ballad "Everybody Is a Star" is an 87. Looking at contemporary pop music, tempos are noticeably higher: Moby's hit ballad "Porcelain" sounds slow at 94, while the average speed of the album it is on, *Play* (which the artist himself considers his "chill-out" record), lies at about 102. A midtempo track is now somewhere around 112, as for example Daft Punk's 1996 "Daftendirekt."

Another major change in rhythm tracks is in their regularity: the tempo difference between different sections of "Lucy in the Sky with Diamonds" is about 4 bpm, the same as on "Us and Them." As Perry Cook and Dan Levitin have pointed out, this is about the minimum tempo difference which can reliably be discerned by a listener in this tempo range (Cook, 931). Perhaps the change to more regular rhythmic patterns has to do with the introduction of drum machines and sequencers; the only explanation that seems reasonable for the overall increase in tempo is a change in fashion. Subtle tempo changes abound on *Dark Side* and contribute vitally to the natural-sounding ebb and flow of its metabolic rate.

In the course of "Us and Them," the music reaches one of its dynamic high points with a couple of full-force ensemble-and-chorus onslaughts which are reminiscent of the grander moments of the Moody Blues. Even so, it is handled well, not overstaying its welcome; the music soon returns to its familiar feel, and after a last verse the scene shifts to track seven, "Any Colour You Like," which comprises delay-enhanced psychedelic synth and guitar solos vaguely reminiscent of Terry Riley's *Rainbow in Curved Air*. David Gilmour never loses touch with the blues roots of rock, though, which is one feature of this track. The penultimate song, "Brain Damage," begins with the return of the massive heartbeat kick drum heard at the beginning of the album, but now overlaid with a chorused guitar riff based on alternating major and minor thirds.

The final track, "Eclipse," is as satisfying a finale as one could wish for. It consists of a song which is purely a chorus refrain, almost every line of the lyric beginning with the words "All that you . . ." Eventually, the music returns to the opening heartbeat drum, and fades to black.

*Dark Side of the Moon* was one of the first seamless rock concept albums, as well as an important step forward in recording technology by virtue of its use of multiple sixteen-track machines and the close-miking of every instrument, particularly percussion. As we have seen, the producers used the studio to bring together disparate elements and create a coherent musical statement. Like the Beatles before them, Pink Floyd used the invisibility of sound and the ability to multitrack in order to make a new kind of performance, one in which the loudspeakers become windows into the music (the image is electronic composer Paul Lansky's). The recorded artifact becomes the compositional container for the many different ensembles, be they of clocks, tom-toms, or live musicians, which make up the musical whole; careful attention to ambiences, EQ settings, compression, and levels become part of the composition, as do the layering and sequencing of tracks.

Production credit was claimed by the band, raising questions as to the producer's role in the studio. In this case, it seems that the group amassed enough know-how from working on previous albums that they felt comfortable taking the helm themselves. In this they were not alone: the Who produced *Who's Next* themselves, while Led Zeppelin albums prominently credited their own guitar god Jimmy Page as producer.

There were and are many arrangements as to who is in charge in the studio. The classic image of the producer in the control room behind the engineer described in chapter 1 had by the early seventies diversified into a multitude of different divisions of labor. One thing was clear: the *sound* of the record was an important element, especially in rock music. How to *get* the right overall sound was a sought-after skill which was constantly evolving and endowed with a certain mystique, as well as the potential for considerable financial reward.

The reality of studio recording is that in most situations, everyone contributes. The musicians come in with their ideas, and in many cases their parts as well, ready to be committed to tape; the engineer will very often come up with ideas; in fact he may himself be the producer. If there is someone with the sole function of producer present, he is of course the main person responsible for the production aspect of a record. Even so, the relationship between artist and producer differs as widely as the CDs in a record store.

As we have seen, the role of the producer has also evolved with technology. When there were only one or two tracks to record onto, and ambience equalled distance of microphone from sound source and size of room, the producer's primary role was to be a talent scout, organizer, and critic. Once three-track was introduced, the idea of balancing elements after a performance came about. Then came rapid advances in tape deck track count, mixing consoles, and effects such as reverb, delay, and compression. As the possibilities proliferated, and hit records were made which featured a distinctive overall sound, the need for people with expertise and ideas grew rapidly. The preferred producer was one with an already established track record.

By the early seventies, there were a number of highly regarded producers known for the distinctive sound of their productions. Building on the advances of Spector and Martin, figures such as Chris Thomas, Glyn Johns (engineer for the Who), and Tony Visconti made reputations for themselves.

## Tony Visconti

Visconti is of particular interest. Known for his "electrification" of former folk singers such as Mark Bolan of T. Rex, and David Bowie, Visconti, a high-school dropout from New York, went on to coproduce classic albums such as Bowie's classic *Heroes* and *Low*, and produced Iggy Pop's *The Idiot.* Working in Berlin's Hansa TonStudio, which featured a huge main recording room which had once been a banquet hall for Nazi leaders, Visconti employed complex, imaginative recording techniques which are still talked about and often imitated. The most obvious, immediately audible feature of his sound is the cavernous room ambience, which flew in the face of the then-prevalent convention of bone-dry studio sound. Although he was not the first to perform radical experiments with room tone (drummer John Bonham's huge sound on *Led Zeppelin IV* is one famous and often-sampled recording), Visconti's tactics were among the more elaborate. He set up drums, bass, and guitar at opposite ends of the banquet-hall-turned-tracking-room (which apparently also featured a live view of the Berlin Wall and machine-gun-toting East German guards), placed mikes at various distances and rolled tape. The result is the hard-rocking, get-up-and-move drum sound heard on hits such as Iggy Pop's 1977 *Lust for Life,* a song made popular again by virtue of its use as the opening track for England's hit film *Trainspotting* (1997).

A more sophisticated technique unique to Visconti was the placement of three microphones at varying distances from vocalist Bowie. Each mike was linked to a gate set to open only when the signal was within a certain range. If Bowie sang softly, only the close mike was operational; at medium volume, the second mike ten feet away kicked in; full-throated singing would open up the third gate. The effect, which can easily be heard on the title track of *Heroes,* is to add more room tone (natural ambience) the more the singer projects. The effect is not the same as simple reverb; the gate cuts off the reverb tail, so that what the listener is left with is a strange, otherworldly quality to the vocal. The contrast is also quite effectively used on the album's fourth track, "Sons of the Silent Age." The verses have a more intimate quality by virtue of the vocal's apparent proximity, while the choruses sound huge and somehow alienated due in no small part to the strange quality of the suddenly distant, yet curiously nonreverberant voice.

The unusual vocal production style is complemented by the elaborate use of outboard gear, which was unusual at the time. There is no use of pre-recorded sound effects. The producers rely on treating standard rock instruments which are actually played. As is evident from tracks such as "Moss Garden" and "Neukoeln" (*Heroes*), as well as "Warszawa" (*Low*), the multifaceted Brian Eno was an important contributor to the sound of both albums. He is credited on *Low* as having contributed "vocals, splinter mini-moog, report ARP, rimmer E.M.I., guitar treatments, chamberlain, piano, synthetics," while the more concise liner note on *Heroes* attributes "synthesizers, keyboards, guitar treatments."

### Brian Eno

Along with Phil Spector and George Martin, Brian Eno is one of the most important producers in the history of pop music. An English art-school graduate who played with the glam-rock band Roxy Music in the early seventies, Eno went on to produce a number of major rock and pop bands, most notably fellow art-school graduates Talking Heads and the Irish group U2. As producer–composer–instrumentalist, he has collaborated with artists such as Robert Fripp, David Bowie, and David Byrne, and also released eighteen albums of his own compositions. Eno is also credited with coining the term "ambient music," and with being one of the founders of the genre.

With the emergence of Eno, the image of the producer moves to a new level. His identification of the studio as a full-fledged musical instrument, a notion influenced by Conny Plank and his circle (see chap. 3), was very important. Beginning in the mid-seventies, his numerous interviews feature discussions of pop music in terms that were altogether new. A fan of "roots" rock as well as reggae, he has pointed out the implications of the techniques employed in the modern recording studio, and he traces his own conceptual lineage as a composer back to John Cage and Erik Satie (Tamm, 19–24). A conceptualist and theoretician as well as an artist, Eno has already influenced generations of pop producers, players, and composers, blurring the distinctions between these roles in the process.

His first production credit is on one of his solo albums, made soon after his split with Roxy Music (there were tensions with singer Bryan Ferry). *Here Come the Warm Jets*, made in 1973, features some of the finest session players working in London at the time. A number of the tracks are listed as collaborations, a way of working which Eno has embraced over the years with great success. This early effort, self-produced and mixed with engineer Chris Thomas (*White Album, Dark Side of the Moon*), already shows very sophisticated production techniques. While the presence of a master such as Thomas couldn't have hurt, Eno is already doing sophisticated things like "ducking" tracks (an effect wherein one signal, usually the vocal, controls the level of the other tracks) and applying unusual ambiences to his otherwise straightforward pop songs.

In his 1979 lecture "The Studio as Compositional Tool," first given at the Kitchen in New York, Eno shared his ideas about recording, composing, and producing in the studio. His talk makes clear that he is already at that time quite aware of the implications of his work, and that he has given thought to the relation in which his work stands to the history of making records. He places the beginning of his involvement as producer–composer at the dawn of the sixteen-track studio, circa 1970:

[The multitrack studio] gave rise to the particular area that I'm involved in: in-studio composition, where you no longer come to the studio with a conception of the finished piece. Instead, you come with actually rather a bare skeleton of the piece, or perhaps with nothing at all. I often start working with no starting point. Once you become familiar with the studio facilities, or even if you're not, actually, you can begin to compose in relation to those facilities. You can begin to think in terms of putting something on, putting something else on, trying this on top of it, and so on, then taking some of the original things off, and seeing what you're left with—actually constructing a piece in the studio. (Eno, 57)

Here the recording studio is effectively a meta-instrument, a way to shape entire compositions. It is score and orchestra rolled into one. Before Eno, pop's sound effects and various delays and distortions were arrangement devices, built around what were essentially still songs in the tradition of Tin Pan Alley and the Brill Building songwriters. Extended psychedelic instrumentals do appear on the seminal Pink Floyd albums discussed above, but it is telling that Pink Floyd was always a live touring act, and that their instrumentals are overtures and intermezzi for the main events, which are at bottom straightforward pop songs.

For Eno, the studio is where composition (not just recording or even arranging) takes place, and what is being made is not a replication or extension of a concert experience, but something altogether different. Earlier pop producers had opened the door to such a conception and had even taken steps toward realizing this idea. But if others had advanced this point of view, it was Eno who embraced it as a long-term project for himself as producer–composer. Beginning with albums such as *Discreet Music* (1975) and *Music for Films* (1978), Eno effectively removed stage performance as a metaphor, taking the listener instead on imaginative sonic journeys unbounded by physical limits. Freed entirely from the pretense of realism, multiple orchestras, synthetic sounds, and other virtual phenomena abound on these and other Eno albums.

Eno is quite aware of how recording technology has reshaped music. In discussing this change, he compares the development of recording to the advent of film:

a point was reached where it became realized that [film] had its own strengths and limitations, and therefore could become a different form through its own rules.

I think that's true of records as well. They've got nothing to do now with performances. It's now possible to make records that have music that was never performed or never could be performed and in fact doesn't exist outside of that record. And if that's the area you work in, then I think you really have to consider that as part of your working philosophy. So for quite a while now I've been thinking that if I make records, I want to think not in terms of evoking a memory of a performance, which never existed in fact, but to think in terms of making a piece of sound which is going to be heard in a type of location, usually someone's house [ . . . ]. I assume my listeners are sitting very comfortably and not expecting to dance. (Quoted in Tamm, 52)

Realism, or the model of replicating a concert experience, just isn't the point anymore. What matters is the sonic experience the record offers, on its own terms, as *sound*. The producer is the director of the aural movie. He has become the composer of the music: in this situation, he is not organizing and optimizing the performances of others; he is himself the artist (although he may of course work with performers and other artists under his direction). This development was really only a matter of time, because once tape-splicing, mixing, and most significantly multitracking were introduced, it became possible not only to reshape the performance after the fact, but also to build up the individual elements on separate tracks without anyone else around. Or, as in the case of *Music for Films* and other Eno albums, one could simply ask a number of people to come in at different times to record on just a couple of cuts. Since one is dealing with only one or two performers at a time in the studio, there are fewer distractions, and one can focus on shaping the performance of that voice in the arrangement. There is of course always a trade-off, in that the spontaneity, the groove found in a good live group is lost; the resulting music may sound artificial, stilted, "cold" if the producer isn't careful. Overproduction was one of the main things punk music was reacting to when it exploded onto the scene in the mid-seventies, just a couple of years after the introduction of the twenty-four-track recorder. Eno avoided being identified as part of the problem, as he was not a fleet-fingered musician of the Rick Wakeman–Keith Emerson school. He was always more of a conceptual thinker than someone given to flashy displays of technique and bombast in general. In 1978, Eno produced the noise-rock compilation *No New York*, as well as the hit band Devo.

As of *Sergeant Pepper*, it had become quite clear to the pop world that performances and records were not necessarily linked. Pop performers reacted to the new situation in a variety of ways. Some groups such as the Beatles or Steely Dan existed only in the studio, while others developed split personalities (a stage act different from their records, as for example the Mothers of Invention ca. 1968) or sought to recreate elaborate studio setups by hauling flatbed trucks full of gear over the road (Pink Floyd).

Eno fell into the first category: his ambient music was not meant to be performed live, as the above quote makes clear, but rather to be heard "in a type of location, usually someone's house."

*Music For Airports: "2/1"*

Perhaps a public space would also be good to listen in, as envisioned by Eno's *Music for Airports*. Released in 1978, this album continued the studio experimentation in the ambient style which Eno had begun in earnest with *Discreet Music*. Like *Discreet Music*, which had rearranged slowed-down fragments of Pachelbel into a new piece of music, *Music for Airports* explores combining and recombining various sound sources in the studio, presenting the listener with a variety of possible levels of listening: one can listen casually, while doing the dishes or reading the paper, or one can focus on the content of the music to the exclusion of all else. Thus the aim of this music is to function on different levels, either as backdrop or as material for focussed listening.

In his excellent monograph on Eno, musicologist Eric Tamm argues that the most successful track on the album is the second, "2/1":

The only sound sources are taped female voices singing single pitches on the syllable "ah," with an absolutely unwavering tone production, for about five seconds per pitch. These sung notes have been electronically treated to give them a soft attack/decay envelope and a slight hiss that accompanies the tone. The pitch material is very limited: seven tones that taken together spell a Db major seventh chord with an added ninth. (Tamm, 137)

Tamm goes on to explain that Eno made tape loops of various lengths and then ran all of them at once (it is not stated whether he did this with eight tape recorders or just layered onto 24-track tape; the latter scenario seems the more probable). The result is a constantly changing sonic tableau in which silence is juxtaposed with shifting melodies and chords which arise from the five-second recordings coming together and pulling apart in many combinations. Tamm (83) has timed the various loops:

Approximate Duration of Pitch-Cycles in "2/1"

| c' | e♭' | f | a♭' | d♭' | f ' | a♭ |
|-----|-----|-----|-----|-----|-----|-----|
| 21" | 17" | 25" | 18" | 31" | 20" | 22" |

Because the music does not follow the structural patterns we are used to hearing, it seems to change on each listening. The patterns of repetition are too complex and take too long to unfold to be perceived as an ordered system; the music's structure is hidden, and what the listener gets instead is a rather pleasant feeling of being somehow suspended in midair. This sensation of air and light is also due to the top-notch recording and processing of the voices, which Tamm describes in the above quote. The effect of the piece is not to push its cleverness into the listener's face, but rather to simply be there; if someone asks, it's quite a cool little scheme, but if he just wants to let the music wash over him, it also works as a backdrop.

Conceptually, Brian Eno is not the first to have conceived of patterns of sound that shift against each other; Terry Riley, Steve Reich, and others had used such concepts before him. What is interesting about Eno is that his career as a "serious" artist unfolds in the same rock and pop magazines and at the same time as his more commercial efforts as a producer. Eno was somehow able to interest a wider public than most composers usually manage to attract. Partly because of his visibility as keyboardist with Roxy Music and David Bowie, and partly because of his articulate interviews and talks, he held the attention of the rock press and the record-buying public. It is also true, however, that Eno's solo records did not sell anywhere near as well as some of the records he produced.

The situation of the producer–composer is not at all simple. If he is to subsist on the fruits of his studio labor, he cannot focus solely on his own music, as this will usually not sustain him economically in the long run. It is almost as though the solo work becomes a kind of professional shingle, hung out to attract clientele. Eno has said as much of *Music for Films*; he even printed licensing info for use by movie producers on the album jacket.

But it may be the producer who makes his own records who is ultimately the most important advancement of the art. In his excellent book *Recording Angel*, Evan Eisenberg considers various categories of producers and comes to an interesting conclusion:

There are [different] kinds of producers. There is the free lancer known for a certain marketable sound, which for a price he will apply like a sauce to the work of a gamy band. There is the technician, neatly doing the bidding of technically ignorant pickers. There is the staff producer in the Motown tradition, cutting raw talent to fit the company last [sic]. There is the survival of Spector's type in the disco producer, whose "groups" may exist in name only, their identities diffused in a pool of studio musicians. And there is the survival of Sam Phillips's type in men like Jerry Wexler of Atlantic, who took Aretha Franklin from hapless Columbia, set her down in a Muscle Shoals studio, and tossed her "I Never Loved a Man (The Way I Love You)" as one tosses a lighted match at petroleum.

But finally it is the artist–producer, the musical creator whose impulse is to create records, who plays the central part in the development of phonography as an art (Eisenberg, 128)

Top modern producers generally have a "sound." It may be varied, it may evolve over time; but pop acts generally seek out a producer with a track record, and that track record typically consists of a certain style of sound which has sold albums. In Eno's case, the hallmarks are the philosophy of writing in the studio, of treating instruments in unusual ways, and the use of chance operations as a compositional tool. Other features one might expect on an Eno project are the liberal use of synths and, depending on what mode he's in, either an ambient or a rock style given to the liberal use of effects. He also has a reputation for the ability to help others do their best, which is a major function of any producer.

The pop producer must be aware of technological as well as musical trends. Eno is no exception, as we have seen already. In the following passage, he offers interesting insights about other aspects of evolving technology and its impact on how we hear pop music:

If you listen to records from the '50s, you'll find that all the melodic information is mixed very loud—your first impression of the piece is of melody—and the rhythmic information is mixed rather quietly. The bass is indistinct, and the bass is only playing the root note of the chord in most cases, adding some resonance. As time goes on you'll find this spectrum, which was very wide, with vocals way up there and the bass way down there, beginning to compress, until at the beginning of funk it is very narrow, indeed. Things are all about equally loud.

Then, from the time of Sly and the Family Stone's *Fresh* album, there's a flip over, where the rhythm instruments, particularly the drums and bass, suddenly become the most important instruments in the mix. A timbral change also takes place. The bass becomes a very defined instrument; by use of amplitude control filters, the bass actually begins to take on a very vocal attack. The bass drum gains a more physical sound, and also has a click to it; generally you'll find that bass drums are equalized very heavily. [The bass drum] becomes the loudest instrument in disco—watch the VU meter while a track is playing, and you'll see the needle peak each time the bass drum hits. (Eno, 50)

Eno's discussion implies that the evolution of recording technology is what has brought about the change in the recorded frequency spectrum. A recorded sound of a pop or rock group which is equally balanced overall means that there were probably multiple mikes to capture the group's individual elements, feed them through a console, and route them to at least three tracks (which could themselves be rebalanced). Moreover, it was the introduction of the compressor, a device which detects peaks in a signal and reduces their levels in a controllable way, which made it possible to make the bass frequencies louder: engineers could now control the amplitude of the most dynamic instruments, the drums and the bass. The widespread changeover to amplified bass was another factor in this development, as was the introduction of the synthesizer with its extended frequency range.

### My Life in the Bush of Ghosts

Putting his theoretical insights into practice, Eno's collaborations with Robert Fripp and especially David Byrne pushed the envelope of what was acceptable as pop, while simultaneously crashing the gates of "serious" music. With records such as the Talking Heads' *Remain in Light* (1980) and the even more radical Eno–Byrne collaboration *My Life in the Bush of Ghosts* (1981), Eno reinforced his image as an experimental electronic producer–composer with a pop audience. The two albums also reflect a keen interest in drums and bass, pushing both to the fore in a manner reminiscent of disco mixes, as mentioned above.

*Remain in Light* was a hit record. A highly successful collaboration, it featured an extended Talking Heads (players included King Crimson guitarist Adrian Belew, as well as vocalists Robert Palmer and Nona Hendryx), with Brian Eno credited as producer, cowriter, and multi-instrumentalist. A very different album from their previous release *Fear of Music* (1977), which was made up of short, guitar-driven rock songs of a relatively conventional nature, *Remain in Light* captured the spotlight for the Talking Heads because of its amalgam of funk, rock, and unusual sounds, as well as its world-beat groove.

*My Life in the Bush of Ghosts* was even more radical in that it was presented to a rock audience without a vocalist or even an identifiable band. The album captured more attention than it might have without the success of *Remain in Light*. The strategy of combining mainstream pop hits with experimental work has served Eno and other modern producer–composers and groups well: a large audience heard *My Life*, and it became a work of long-term importance, its sampled-vocals-recontextualized approach echoing through many a current electronica album.

The recording techniques employed on *My Life* were state-of-the-art at the time, but twenty years of technological development is a long while. The main difference between then and now, immediately audible upon listening to the first track, is that the drums are played by a real live drummer throughout. Drum machines weren't around yet; the first commercial drum machine which employed sampled sounds, the Linn LM-1, was released in 1979, at the same time *My Life* was being recorded. Only five hundred were made. The ever-popular Linn Drum, a staple of eighties disco and much pop music, wasn't released until 1982.

Another difference is the total absence of computers in the recording studio. At the time, practically every professional pop project was recorded to twenty-four-track machines. Samplers also did not yet exist; "sampling" meant transferring a recording on tape or off a record onto the two-inch tape on the twenty-four-track. The first commercially available sampler, the Fairlight CMI, also came onto the market in 1979, while *My Life* was being recorded.

This album thus sits on the cusp of a variety of changes and developments in popular music in more ways than one. Eno and especially Byrne were aware of the new art emerging from the South Bronx: graffiti, break dancing, and rap (Hager, 40). Byrne was also quite interested in types of music from around the world, especially African drumming. In collaboration with Eno, he set about popularizing what eventually became the genre now called world music. Also of interest in this context is the presence of Bill Laswell on bass (on track one of *My Life*), who would himself later become a leading producer of the world style.

*My Life in the Bush of Ghosts* features a variety of non-Western sounds, both sampled and recorded. All of the vocal performances are external sounds "flown in" (a studio term for syncing-up sounds not recorded in the studio with the tracks of a session) from a variety of sources. This was quite unusual for a pop album in 1980: a pop producer and a star singer make an album on which the singer's voice isn't heard at all. Thus the credits list a Lebanese mountain singer, an Egyptian pop singer, Algerian Muslims chanting the Qu'ran; on other tracks, an unidentified radio evangelist from San Francisco, an "indignant" radio host, and a "smooth politician" hold forth over the music's infectious pulse. As if to further underline its diversity, the album credits five studios and ten engineers as having contributed to the project (not including the mastering house and engineer).

The fourth track of *My Life*, entitled "Help Me Somebody," merits especially close listening. It sounds as though it could have been recorded last week, but like the rest of the record, it was actually recorded about two decades before this writing. Like the album as a whole, the music is located in the rock universe by virtue of its rhythm tracks, but these are not rock songs—or at least not conventional rock songs. An interesting feature of

"Help Me Somebody" is that there actually is a song structure, but the upper range is mixed low relative to the steady pulse of the rhythm section. The result of the mix is an effect of minimizing the importance of the changes over the groove. It is possible to listen casually to the track several times before one even realizes the structure built over the drums and bass.

This is part of Eno's plan. In the days before sound recording, he says,

The piece disappeared when it was finished, so it was something that only existed in time. The effect of recording is that it takes music out of the time dimension and puts it into the space dimension. As soon as you do that, you're in a position of being able to listen again and again to a performance, to become familiar with details you most certainly had missed the first time through, and to become very fond of details that weren't intended by the composer or the musicians. The effect of this on the composer is that he can think of supplying material that would actually be too subtle for a first listening. (Eno, 92)

"Help Me Somebody" does just that, providing a surface texture consisting of congas, drum kit, bass, and two muted rhythm guitars, which provides a fast (146 bpm) and steady pulse throughout the track. As the cut fades up, birds are heard along with the first impression of fast conga playing. They are dutifully credited in the liner notes as "Rooks on 4." On close listening, it turns out that there is also a kit playing, but with a bandpass filter set to allow only a portion of the midrange to be audible. Once the fade-in is complete and the first vocal sample has been heard, the drums suddenly switch to full frequency response (0:26), and seem to jump-start the bass and rhythm guitar. The vocal track (culled from a recording of one "Reverend Paul Morton, broadcast sermon, New Orleans, June 1980" [Eno and Byrne]) then makes its first appearance. The next significant event is a smooth little guitar riff beginning at 0:52. It's a welcome addition to the piece, and is played only a couple of times before fragmenting into the mix. This is another compositional device used throughout this track: riffs and sounds will appear, make a brief statement, and then reappear only as traces or fragments of themselves, usually mixed in with other traces of similar provenance. Once the (again low-in-the-mix) metal percussion break happens, for example, it returns only as an accent, and combined with other fragments (2:10, 2.23).

There are other subtle touches as well. Close attention to the kick drum reveals beautiful delay work during the metal percussion break, effectively writing a new part for the kick using effects. There is also a skillful interplay between guitar and sampled vocal beginning at 2:28, which is all the more impressive when one considers that slipping tracks has only become widespread since the advent of nonlinear editing (ca. 1989). Again, these samples had to be transferred, or flown in, from one tape machine to another, no easy task when one was attempting to sync up spoken word fragments with instrumental tracks. Similarly, there is considerable playing skill involved in sustaining metronomically precise conga and drum patterns at the rate of 146 bpm. The groove is interesting, because although the track is obviously referencing funk (the placement of the kick is on one and the last sixteenth of three), it also has a machinelike metronomic pulse reminiscent of the robotic feel of early drum boxes such as the Roland 808. As we shall see in the next chapter, the rise of synthesis and sampling technology was having an effect on the grooves of many flavors of pop; whole new styles came about as the result of these new devices.

The music discussed in this chapter so far is indicative of the emergence of a new role for the pop producer as someone who does not merely oversee the recordings of others but is himself a recording artist on the project. This is of course not a universal change; there are still plenty of "old-school" producers working today. They are necessary to the making of many recordings, and will continue to thrive as long as nonelectronic music is being committed to tape and CD. What changes beginning with the late seventies is that there now exist producers who have substantial careers as creators (composers) of their own work and who apply elements of their own sound to the projects of others. It is interesting to hear their various sensibilities in the work of those they produce—aside from Eno and the Talking Heads–Bowie–U2, the Laswell "world" sound on Swans' *The Burning World* and Ginger Baker's *Horses and Trees*, for example, or the similarities in sound between Trent Reznor's (Nine Inch Nails') own effort *The Downward Spiral* and his production on Marilyn Manson's *AntiChrist Superstar.*

## Bill Laswell

Bill Laswell is a versatile producer and bassist. I worked with him in 1988–1989 on the Swans album *The Burning World* as drummer for the band. At the time, Laswell had gained prominence in the pop world for having coproduced Mick Jagger's solo album *She's the Boss* (1985), and for having cowritten Herbie Hancock's smash hit "Rockit" (1983), which one critic has described as "the song most often associated with the break-dancing movement of the mid-80s" (Ross, 3). Originally from Detroit, where he developed his funk style, Laswell moved his operation to New York around 1979 and soon became a leading figure in the Downtown underground punk-jazz scene. With his group, Material, Laswell worked with an exceptionally wide variety of musicians, recording with everyone from jazz saxophone legend Archie Shepp to the then-unknown Whitney Houston.

Once Laswell had a hit with Hancock and worked with Jagger, both rock bands and funk acts were eager to work with him. Around 1986, Laswell also adopted a "world" sound which was heard consistently across a number of his productions, from former Cream drummer Ginger Baker's solo album *Horses and Trees* to Manu Dibango's *Afrijazzy*. His own music bears a strong resemblance in terms of overall sound to some of the albums he has produced. *Hear No Evil*, for example, features the same core group of players which appears on a number of Laswell-produced CDs, including *The Burning World* (MCA, 1989).

Bill Laswell is an unusually prolific producer. He has made literally hundreds of records in the last twenty years. His process of development has been additive rather than serial. At the time Swans worked with him, Laswell was also making records with funk legend Bootsy Collins (Parliament/Funkadelic), baritone sax wildman and free-jazz legend Peter Broetzmann, punk-jazzers the Golden Palominoes, and his own Material.

His production methods are very flexible; in fact, Laswell comes across as being something of a chameleon. Not everything he does bears his recognizable thumbprint. On some albums, his presence is barely felt. Yet in the case of the Ginger Baker album, Laswell has said that he simply took rhythm tracks played by Baker at the time of the PiL *Album* sessions and built the music up himself, with the aid of the usual crew of diverse and talented musicians. On Hancock's "Rockit," Laswell's bass plays a prominent part, and, as on many of his projects, Laswell is credited as cowriter.

Overall, the impression one is left with after listening to a number of Laswell productions is that he is extremely adaptable. Unlike a "personality" producer such as Brian Eno, Laswell seems to make records which fall

all along the spectrum of production methods. On one end, he acts as an old-school producer, content to bring people together and stay in the background, while on the other, he takes over almost completely, coming very close to simply putting someone else's name on what is essentially his music. He is different from someone like Eno, who puts his stamp on everything he is associated with. Eno is the first major producer with an authorial voice which is consistently present throughout his career. In this respect, he is most like a film director, whereas Laswell's polymorphous production activities tend to dilute his identity as a composer.

## The Making of Swans' *The Burning World*
In the case of the Swans recording project I worked on, Laswell's influence was strongly felt; he had a major impact on the sound of the CD we made together. Swans singer–songwriter Michael Gira had actively sought a producer with a "sound" to work with on this first major-label effort by the band. He had approached Brian Eno but was turned down. Gira persisted in seeking out a producer who would bring a new, somewhat brighter sound than the band was known for to its next album; eventually he was introduced to Laswell, and after a period of discussion, they agreed to work together. Laswell seemed to make sense for Swans for a number of reasons. Aside from his strong track record of record sales, he brought to the band both rock and world music credits. In early 1988, I was asked to join Swans. The group consisted of Michael Gira, vocals; Norman Westberg, guitar; Algis Kyzis, bass; Jarboe, keyboards and additional vocals; and myself on drums.

Like many bandleaders, Gira's approach was to use an extended tour to perfect new material, then go back to New York and record an album. During the writing and rehearsing stage, Michael would come up with simple chords and riffs on his acoustic guitar, which he would then play for one or more members of the band, often singing as well. If he wasn't sure about what he wanted, he would arrange to meet individually with one or more players for an hour or so of acoustic, low-volume prerehearsal. The rehearsals were an arduous process. We rehearsed four hours a day, five days a week, for almost three months before we went on tour. In 1988, Swans played in nine countries and thirteen states, sometimes in front of thousands of people. The band was nearing the height of its success.

Three months after the tour ended, we went into Platinum Island Studio under the auspices of MCA Records to begin recording. By the time the record was finished in early March of 1989, it had taken on a completely different sound than the music we had arranged and played on tour.

The studio recording process was one in which the Swans sound was fed into and processed by the Laswell production machine, with the final result being a disc which is half Swans and half Material (Laswell's production company and erstwhile band). The process of transforming the music from our stage sound and arrangements into a sophisticated studio product offers insight into the way a major-label album aiming for chart success was produced in the late eighties.

The initial phase involved meeting in the Swans rehearsal studio in the East Village for preproduction. The first thing that happened was that Laswell came to the studio by himself and simply listened to the band run through the set. At that meeting, Gira gave him a two-page document outlining his ideas for the production of each song. "Let It Come Down," for example, was explained as being based on a novel by Paul Bowles of the same title, and the production memo called for strings and percussion to be added to the existing guitar–bass–drums arrangement.

A couple of days later, Laswell showed up again, this time with his right-hand man, guitarist Nicky Skopelitis. They brought along a Fairlight CMI, which was already obsolete technology at the time, but proved quite useful nevertheless. Laswell explained that we would get a better feel on the drums if I played not just to a click track but to a complementary groove. For example, if a part were made up of snare, kick, and hi-hat, they would have me play the part, and Skopelitis and I would come up with a complementary voice on virtual toms, which would then be programmed into the Fairlight. When the time came to lay down the tracks in the studio, my headphone feed was not merely a click, but rather the arrangement we had come up with on the Fairlight. I found that this helped me create the basic rhythmic feel for the record, as I was interacting with something more than a simple succession of eighth-note ticks.

One of the major reasons for recording each instrument separately was that this gave the producer the freedom to replace parts at will. In the case of "Saved," I was surprised to come into the studio one day and find that my beat had been surgically altered. The snare on four had been removed in the second and fourth bar of the chorus. On "See No More," I was now sharing the role of drummer with Ayib Djeng, the well-known West African drummer. It is worth noting here that as with Eno and Byrne's *My Life in the Bush of Ghosts*, the activity of replacing and moving things around at the quarter-note level was quite labor-intensive and required a great deal of skill. Laswell surrounded himself with devoted professionals who would go out on a limb to achieve what he asked for technically; he himself did not touch the board.

The drums were recorded in the large room at Platinum Island, which is large enough to fit a symphony orchestra. The only people present in the studio during the day were Laswell and Robert Musso, the engineer. In the evenings we would listen to the day's work and talk about it with the whole band present. The atmosphere was relaxed at this point, as everyone had money in the bank and the music was sounding good.

Guitars were next. Norman Westberg fairly ripped through his parts, without any hesitation or perfectionism. His sound was simply great, and the drums and guitars together made a glorious noise through the huge monitors in the control room, which was vaguely reminiscent of the space station in the movie *2001: A Space Odyssey*.

The bass parts were recorded after the guitars. This was because the band had a new bassist, Jason Asnes. Jason had been hired to replace Algis Kyzis, who had quit the band after the tour. Laswell reasoned that because Norman had been with Swans for years and knew he material inside out, his parts would serve as the basis for Jason's bass. However, this second bassist didn't last either, so Bill played the remaining bass parts himself, getting them done in one night.

Around this time, the first of many session players showed up. His name was Jeff Bova, keyboardist, and he arrived with enough road cases to fill most of the studio. Jarboe, who played keyboards with the band live, did not consider herself a keyboard player, so Bova was brought in to play her parts. He had every high-end synth known to man at that time. His parts on the CD were an interesting improvement over the parts Jarboe had played on tour. They weren't more complicated, they just sounded "right" (I shall discuss one track in detail later on). Like engineer Bob Musso, Jeff was part of Laswell's regular crew, so things went fast and smooth.

At this point, we were about three weeks into the project. So far, things had gone more or less the way one would expect them to go in any small studio. Aside from some clever touches like the groove-as-click concept, the only difference was one of scale, and that there was someone in the studio at all times whose sole function was keeping the project on track, listening and guiding the proceedings.

The difference between project recording and big-studio techniques became evident with the next phase of production, which was the first part of manipulating the recording. Effects specialist Bruce Calder was brought in, again with tons of equipment. He immediately set about performing a host of interesting operations on the tracks we had recorded up to that point. The guitar was given delay, but not by means of a stompbox (electronic footpedal); rather, a Revox varispeed recorder was employed for extra "warmth" (harmonic distortion due to slight tape saturation). The signal was fed into the record head and mixed with the signal on the play head; the varispeed control was used to obtain the right amount of delay. The snare track was soloed and fed through a JBL 4312 monitor in the recording room, with an EV RE-20 microphone placed at the other end of the room to pick up the sound. This was done in order to get a separate, natural-sounding reverb on the snare alone.

Laswell maintained a respectful distance from the workings of the effects person. Essentially, that engineer was doing pre-effects and premixing. The idea was to make the final mixdown easier and provide the vocalists with a great-sounding track to sing over.

The next part of the project was the highly sensitive phase of recording the vocals. The work was to be done in Brooklyn, at Martin Bisi's studio (Laswell later built his own, also in Brooklyn). Many rock vocalists take great care in recording their voice. Often, attention is paid to the inflection of each word, syllable by syllable. Gira and Jarboe were no exception; it took six weeks to record about thirty minutes of singing.

The last two stages of the recording were sweetening and mixing. Jazz composer and arranger Karl Berger was brought in to write string charts and play vibes for some of the material, and Laswell's circle of musicians laid down a host of overdubs. This the point at which the record stopped sounding like Swans, and became a hybrid of the Laswell sound and the music we had played on tour.

"See No More," the penultimate track of the album, can serve as a concrete example of the processes I have outlined in a general way. There exists good documentation of the evolution of this track: the original "feel" click tracks described above, the CD version, and a live tape from a Swans performance in Amsterdam in the spring of 1988.

The structure of "See No More" is very simple. In the live version, it consists of:

intro (cymbals and drone)

chorus (11 × 4-bar riff)

verse (8 intro + 8 × 4 [+2])

chorus (6 × 4)

verse (4 intro + 12 × 4 [+2])

verse riff expansion (8 × 4-bar riff)

outro (8 bars)

Many aspects of the song were altered for the studio version. The structure was left essentially intact, if somewhat abbreviated:

(no intro)

chorus (8 × 4-bar riff)

verse (4 intro + 8 × 4 [+2])

chorus (6 × 4)

verse (4 × 4)

chorus (4 × 4)

outro (8 bars)

The most obvious effect of these alterations is that the studio recording of the song is about one-third shorter than the live version (live: 8:24; studio: 5:31). There are also many other changes: there is a pronounced contrast in rhythmic feel between the two versions, and there's a considerable difference in tempi between the choruses and the verses of the live version. The chorus as performed live clocks in at about 126 bpm, while the verse varies between 98 and 104 (accelerandi occuring as a phrase-shaping device). Laswell suggested an interesting change for the studio recording: he did away with the tempo changes and made everything 110.

Perhaps the most interesting change as far as the basic character of the song was the feel. Where the original (live) chorus had a very four-square, brute-force insistence on each quarter-note throughout the phrase, the reworked feel employs graceful upbeats on the last 16th-note of the one and three, and turns into more of a gallop than a goose-step. The live version of the drums was closely tied to the riff, whereas in the studio version, the rhythm tracks provide a kind of counterpoint to the melody.

Another change in the studio version is the orchestration of the track. In addition to the usual guitars, drums, and bass, Laswell added West African drums (played by Ayib Dyeng), which can be heard on the left channel of the recording, opposite my toms on the right. The final chorus features bazouki, a mandolin-like instrument played by Skopelitis. These devices serve to sweeten the track's sound, softening the forbidding quality heard on the "Live in Amsterdam" version.

On that live recording, the vocalist's mannerisms seem excessive, and probably reflect the excitement of playing to a packed hall of well over one thousand spectators at the Paradiso, simulcast live on Dutch National Radio. The excitement of live performance may also account for the speed of the choruses. The recorded version's vocals are more carefully crafted, and more convincing. Another change which contributes to the more polished, musically developed sound of the album version of "See No More" is the simple yet effective keyboard ornament which seems to grow out of the drum and guitar figures each phrase. By playing a figure which repeats the fourth, fifth, and sixth scale degrees, keyboardist Jeff Bova manages to minimize the jarring effect of the B flat heard in the first part of the melody.

"See No More" has a built-in awkwardness due to the insistent character of the main riff. To understand why this is, it is helpful to look back on Swans' early music, which dates from the group's beginnings in 1982 to about 1986. This earlier, more brutally loud and raw-sounding version of the band was a famous early exponent of noise-rock (others include Sonic Youth, Elliott Sharp, and DNA). Swans albums such as *Cop* and *Holy Money* relied on painfully slow, grinding two-bar loops, overlaid with the agonized voice of Gira singing lyrics which make the already dark "See No More" seem light in comparison. The early Swans material was about simple, brutal, hard-hitting drums, combined with samples which often evoke the image of large, ominous machines or omnivorous monsters. Thick slabs of guitars and bass were added to complete the monumental, oppressive, yet somehow also perversely erotic effect of the music. The early Swans sound helped shape the style known as "Industrial," and is related to bands such as Ministry, Skinny Puppy, and Einsturzende Neubauten.

By 1988, Swans had made *Children of God*, a successful, rather tuneful double album, and Michael Gira was sounding more like a cross between Jim Morrison and Joy Division's Ian Curtis than the growling beast heard on albums such as *Cop*. Yet the songwriter Gira did not leave the extreme performance-art aspect behind in one fell swoop and become a musician in the singer–songwriter tradition as exemplified by, say, Bob Dylan or Woody Guthrie. Insistent repetition continues to be a theme in Gira's music throughout the nineties; songs usually consist essentially of two alternating riffs, or even of just one phrase, repeated throughout the entire song. *The Burning World* is a transitional album; hence the sometimes jarring and not always effective transitions between sections of songs such as "See No More" and "Jane Mary."

Gira and his various collaborators employed effective production techniques to make more successful projects out of albums such as *White Light from the Mouth of Infinity* (1991) and *Love of Life* (1992). "Failure," a song from the former album, features an unusual equalization on the main vocal to give it a distinctive character; other songs use various devices such as loops of children's voices and real-world sounds to enhance the overall ambience of the production.

### Trent Reznor

Another important producer–composer is Trent Reznor, who catapulted to fame with his 1994 release as Nine Inch Nails entitled *The Downward Spiral*.

Reznor is similar to Laswell in that he wears a number of different hats. He also brings a great deal of skill and savvy to everything he does, be it as a songwriter, composer, arranger, engineer, guitarist, keyboardist, or producer. In the ten years since he first appeared on the scene with his album *Pretty Hate Machine*, Reznor has not only built a major career fronting and writing/recording all material for Nine Inch Nails, he has also been involved as a producer for other major acts (e.g., David Bowie, as well as Florida metal/industrial act Marilyn Manson). In addition, he runs his own successful label, Nothing Records (Nine Inch Nails, Manson, and English electronica acts Autechre and Squarepusher, among others).

Like Eno, Reznor has a signature sound: his is hard, abrasive, grating, yet also connected to the American songwriting tradition, however obliquely. Reznor's success seems partly due to his ability to both break with the traditions of rock and continue them at the same time. He has found the right mix for contemporary pop: iconoclastic at first glance, very high-tech, but with an underlying romantic, tonal sensibility which ultimately carries the music.

Like most other pop writers, Reznor uses the studio as his note paper. The studio tape is his score; the production is an integral part of the composition. Moreover, the studio is the orchestra; by this I mean that the musical instruments as well as the assorted noises that make up the final mix are what make the music happen. The orchestration, or studio arrangement, is an integral part of every song.

This does not mean that Reznor has no collaborators, or that what he does is conceptually analogous to the methodology of nineteenth-century composers. Clearly, when people work in the studio, there is often collaboration. But I do not believe that everyone present in the studio is what Albin Zak has called a "recordist," and hence automatically involved in the process of production. There are clearly producers who have a vision, an overall aesthetic. Reznor is one of these figures. A drummer's offer to play with the snares off, for example, makes him no more of a producer than a violinist suggesting to Brahms that a passage could be played *con sordino* makes him a co-composer.

Another important feature of Reznor's art is that he was one of the first figures of the nineties to bring the new digital technology into the consciousness of mainstream America. He built on what Eno and others had done before him, but went one step further by mounting a frontal assault on the "no keyboards" aesthetic of grunge and appealing to its large and young audience. "Maybe keyboards are considered unfashionable these days, but I don't give a fuck," he intoned with typical directness at the beginning of a 1994 interview in *Keyboard* magazine. "I think keyboards have been given a bad rap in rock music [ . . . ]. The Pearl Jams and whoever, that's not what I'm about. I like keyboards, I like technology. This is who I am" (quoted in Rule, 4).

One link that leaps off the page in this interview is the direct association between the terms "keyboard" and "technology." Reznor is making a connection which is logical enough if one looks at the production of musical instruments in the contemporary industrialized world. Keyboards are closely associated with synth engines, computers, indeed all things in music which are high-tech and have been completely digitized, replacing to a large extent the Wurlitzers, Rhodes, and analog synths of yore (Moog, ARP, Oberheim, etc.). Yet guitars have remained almost exclusively analog.

Reznor has paradoxically embraced all things digital, while at the same time recuperating the digital keyboard and the technology it is associated with as directly and instinctively connected to the performer. This primary and primal connection is essential to rock, and again explains Reznor's impact: there is a ghost in the machine, and he's not very nice. But like most demons, he is a fallen angel, and speaks to his audience in a very direct language which appeals to conflicted human emotions. And these themes are expressed with a new voice: Reznor's work uses the language of contemporary, wired America. There is a direct connection between his message of rage and alienation, which is simultaneously a bid for redemption from the cold, soulless aspect of technology, and the way Reznor chooses to produce his music.

*The Downward Spiral* was one of the first major albums to be recorded and mixed in a home or project studio. With the development of affordable, portable technology which began in the late seventies, it has become increasingly possible to create high-quality masters which are indistinguishable from recordings made at major studios. The advantages of the project studio include unlimited time, equipment tailored to the music, a more relaxed atmosphere, and total control over the product. If there is an outside engineer, he is working closely with the artist and is usually in the artist's space.

Not everything is simply better and easier working in a project studio, however. One has to make everything work without the benefit of a support staff, and the new machines, especially ones that involve sophisticated software, tend to be barely ready for market. Sometimes "features" (i.e., bugs) that should have been eradicated in beta-testing can pop up during a session, bringing it to a grinding halt. Reznor is very up-front about the potential pitfalls of getting closely involved with music machines:

Keyboard:   Were there many gear snafus or other problems associated with working at home?

TR:   There were a couple of pieces of gear crucial to the way the studio worked at the house. One of them was a Time-Line synchronizer. It syncs the two Studers, Pro Tools, and everything. To be honest, it didn't work. Ten times a day we'd have to turn it off, unplug every cable, plug 'em back in, turn it back on, call the company, and say "Guess what? It doesn't work." There were many times when I thought, "Am I the only person in the world who's ever tried to hook these two pieces of gear, that they say work together, together?" So between that, and the terrible automation on the Amek board we had, things would grind to a halt. I cannot tolerate equipment fucking up when you're trying to write a song, when you're on a roll.

One danger of having a full studio in your house is: What do you focus on? I could spend, and have spent, a month just sampling things. So now, when it comes time to pull up a drum bank, it's all cool sounds that I've created, rather than leftover from things I've used before. We spent a lot of time sampling and processing the sounds through different things. That way, when the actual writing and arranging moment came . . . when you went to reach for that bank of sounds, they were inspiring, instead of, "Fuck, I'm in the middle of writing a song, but I really should spend a couple of days getting all new Oberheim sounds." (Rule, 5–6)

One thing that stands out is how quickly the technology of music has evolved just in the last six years. Today hardly anyone bothers with elaborate sync setups anymore; things are recorded into digital audio workstations such as ProTools or Logic, and then copied to twenty-four tracks for the final mix. Even that is changing, with an increasing number of projects simply being kept in the digital domain.

The *Keyboard* interview brings out just how much of a hands-on producer Reznor is. His level of technical skill is reflected in the originality of his music: his instrument is the studio, and he knows how to play it without relying on intermediaries. It is one thing to have assistants perform rote tasks such as cataloging samples or indexing DATs; it is quite another matter to tell the engineer "make it sound fatter," or "make it sound more green," and then rely on his skills to make it sound right.

As we have seen, the way the drums are arranged and recorded is an element of the sound which is often crucial to the feel of the music. Reznor is no exception. His take on recording and sampling drums is particularly intriguing, because it involves a peculiar *trompe l'oeil* effect:

Everything was programmed. My idea of a drum is a button on a machine. When I hear a real drum kit . . . when someone hits a kick drum, it doesn't sound to me like what I think a kick drum is. Any time I've been faced with, "Let's try miking up the drums," well, you put a mike up close, you put another one here, 300 mikes, gates, bullshit, overheads, bring 'em up and listen to it and it doesn't sound at all like it did in the room. It sounds like a "record sounding drum kit." It doesn't sound like being in the room with live ringy drums. You read these interviews where producers will say, "It sounds like you're in the room with the band." No it doesn't. Nirvana's record doesn't sound like you're in the room with them. It might sound sloppy, and it sounds interesting, but it's not what it sounds like in the room, to me, anyway.

So we were experimenting with just two mikes, PZMs usually. We ended up taking a drum kit into about 25 different rooms. [We] put mikes at about the same distance from the drums, then hit each drum at several velocities and recorded them on a DAT machine. Then we sampled them all with velocity splitting on the Akai 1100 samplers. I noticed that when you sat down and played those on a keyboard, they sounded exactly the same way they did in the room: shitty, ringy, you know. And that, in itself, lent a strange, unexpected vibe to the thing. So on a few songs, we used that. I made the drum programming very rigid, so that maybe someone will listen to that and think: "is that a drum machine? Nah, can't be. No machine sounds *that* shitty." I like the idea of hearing a record and thinking, "That's guitar, bass, and drums," and then, on further inspection, "Wait a second, that's not what it appears to be." (Ibid., 7)

This blurring of the distinction between recordings of real instruments and recordings of recordings is an interesting game, and it fits right in with Reznor's overall strategy of playing a kind of push-and-pull with the two poles of record production, the illusion of reality ("this was played by real people in a real setting") and the reality of illusion ("this doesn't exist in the real world; we're making our own universe").

### "Mr. Self Destruct" v. "Irresponsible Hate Anthem"

It is impossible to separate Reznor the producer from Reznor the artist. They are one and the same. His functions as producer and composer overlap: his orchestra is the studio. The instrumental lines move not only tonally, but timbrally; sometimes in a linear fashion, more often disjointed, yet with a peculiarly compelling logic which resonates with our time. Reznor's work as a producer is clearly an extension of his personality as a composer. One need only compare the first song from *The Downward Spiral*, "Mr. Self Destruct," with the first cut form the 1997 Reznor-produced Marilyn Manson album *AntiChrist Superstar*, "Irresponsible Hate Anthem," to see the close connection between the different facets of Reznor's art.

Both songs are tonally and structurally simple, and are very similar in production style. "Mr. Self Destruct" is based on a D pentatonic scale and alternates between the first and third scale degrees, while "Irresponsible Hate Anthem" uses five notes of an F-sharp blues scale throughout. In terms of both structure and production, the Nine Inch Nails song is better crafted; but the similarities in production and overall color and mood of the two works are what's interesting. It's almost as though Reznor grafted (or transposed, to use a more musical term) the elements that went into the recording of "Mr. Self Destruct" onto Manson's "Hate Anthem." It also may be that he saves his very best ideas for his own records; or perhaps Marilyn Manson's sensibility is simply less sophisticated than Reznor's.

Both songs also open with nonmusical noise, but while Manson's is simply crowd noise, Reznor's is not quite identifiable. The opening soundscape of "Mr. Self Destruct" sounds like someone being flattened by a knockout punch or some similarly unpleasant scene. As one listens more, it turns out to be a loop, which is gradually speeded up until it erupts into the manic first verse of the song. Once the music begins, one hears instantly the sort of ambiguity between real and sequenced "performance" Reznor is talking about. The bass line sounds like it is being played by a real player, but it is just a little too fast, a little too precise and mechanical for that to be the case. The drums are also unusual and not trivial to analyze, although as one listens to the song, it goes by easily. The ear doesn't stop and say "wait, that's strange," it simply gets swept up in the forward motion and intensity of the song. On close listening, there turns out to be a kick on every quarter-note, with an electronic sound containing white noise, probably a sample of some sort, providing a regular, eighth-note shuffle feel in the upper or snare-and-hat register. There is also some sort of drumlike sound playing impossible-to-play 32nd-notes in perfect time with the bass, which makes for an original-sounding texture.

Underneath the metallic surface of the music, Reznor's gift for the simple yet compelling musical statement is clear. Like most good pop, this song moves along with the sort of inevitability one also associates with the great classical composers. Each new part of the song seems to follow logically and inexorably from the previous one. Simple devices are at work, but all of these simple devices are in all the right places in all the right proportions. For example, there is an elementary symmetry between Reznor's descending verse melody and the consequent rejoinder of the chorus. It's an effective call and response, which is followed by an eerily quiet bridge featuring a bass line, again in D minor. The build in this section is provided by the vocal, which begins with a whisper and gradually builds to a threatening mumble. There's an interesting combination of elements at work here; the voice ramps up over a steady bass, which does not grow louder with the voice. The dissociation between these two elements is very effective. When the chorus returns, its impact is explosive. The contrast between the extended, uncluttered bridge and the in-the-red chorus makes for a ferocious impact. The lyrics are also powerful, and work in tandem with the music to paint a disturbing picture of alienation, rage, and destruction.

The quality of the vocals is also interesting. They seem to want to jump out from the speakers and grab one by the throat. The production is intriguing from a timbral point of view, because there is some sort of distortion device at work, possibly something as simple as a guitar distortion stompbox such as a Roland GTR-70 or a Zoom 9100. Varying levels of distortion and bandpass equalization are used to create different registers and qualities for the voice; in the choruses, for example, there is an alternation between the main voice and another, more distorted voice which is mixed lower. The effect is claustrophobia, and it comes across as a musically compelling construct. If one has an urge to leave the room, it is because it is frightening, not because it's poorly made.

There is one element of Reznor's sound on *The Downward Spiral* which is a bit disappointing to my ear. It's the ubiquitous "line-in" quality of the guitars and distortions overall. It sounds as though in many cases, cheap line devices were plugged into expensive, high-end equipment (Amek board and Studer 24-track 2-inch recorders), which faithfully captured their unresponsive, undifferentiated brand of harmonic distortion. Anyone who has played in a hard-rock band or listened to the recordings of Jimi Hendrix knows the difference in sound between a screaming Marshall stack full of vacuum tubes gone critical and the widely available approximations thereof which employ digital circuits. It just doesn't sound the same. Getting a good distorted sound is an art; one example of a truly superb sound of this type is Prong's *Third Planet from the Sun*. The Manson album has a similar guitar sound to *The Downward Spiral*, so this was apparently the sound Reznor favored at the time. By the time of *The Fragile* (1998), Reznor has developed a far more subtle yet full-sounding guitar timbre.

Distortion is an art. Robert Walser writes about it in some detail in his treatise on heavy metal. He begins by noting that the human voice produces distortion through excessive power (screaming), and goes on to say that

the distorted guitar signal is expanded in both directions: the higher harmonics produced by distortion add brilliance and edge (and what guitarists sometimes call "presence") to the sound, and the resultant tones produced by the interval combinations of power chords create additional low frequencies, adding weight to the sound. (Walser, 43)

The last section of "Mr. Self Destruct" delivers on the promise of the song's title. As the chorus cycles through its big F–G–A–D, it is gradually engulfed by a tidal wave of pink noise which ultimately obliterates everything. Well, almost everything: once the wave has come to dominate the mix, there is a sudden cut to a twisted-sounding guitar loop suggestive of a broken machine emitting demented screeches. This three-second loop is repeated about fifteen times before the music suddenly switches gears again, launching into the second, very different cut of the album.

The music of Marilyn Manson is very different from Reznor's. Yet the surface connection between the two songs is quite strong, and is due to Reznor's production. Like "Mr. Self Destruct," Manson's "Irresponsible Hate Anthem" is based on a very simple riff, as noted above. The F-sharp blues scale is boxed into a four-square rhythmic grid, and there isn't much subtlety behind anything here: it's just meant to be hard-hitting rock, and that's what it sounds like. Aside from the similarities already mentioned (the opening and ending noise segments, the guitar sound, the simple, obsessive riffs), there is also the similar vocal sound and performance, which is dry, distorted, and ranges from whispering to over-the-top screaming. The placement of the kick drum on every quarter-note is another similarity, although it does alternate with a more differentiated beat in the Manson. Indeed, the drums here are clearly played by a person, not a machine. This human presence is generally more felt, and is the major underlying difference between the tracks: the bass is also straightforward, its obsessive blues-turned-scary-metal riff faithfully pounded out by bassist Twiggy Ramirez, and there is the overall feeling that although the tracks have been polished and tightened, they are based on real-time performances, rather than sequencers. Marilyn Manson sounds like a slickly produced rock band. They have taken on some of the timbral veneer of Nine Inch Nails for this record, but underneath it all they're still a more conventional stage act.

The Manson record, then, is based on the concept of live band performance. There's no question of machines v. live players here: the liner notes indicate that the cut was "performed live, February 14, 1997" (Manson). What is left unclear is whether this was a live performance in the studio or onstage somewhere. If it was played before a live audience, it's curious that no performance venue or city is indicated.

Even Nine Inch Nails is a more traditional act than the types of music we'll look at in the next and final chapter. Although Reznor does create his records in the studio using any device that's useful to him, he also sees himself as a performing artist in the rock hero (or antihero: Jim Morrison) tradition, and subscribes to the live performance model of music. Like other major acts such as Beck, who create in the studio using the full range of techniques available to them and then assemble a live band for touring and TV appearances, Reznor has not dispensed with the traditional idea of the band. Or maybe it is more accurate to say that these artists and producers (Gira, Reznor, Beck) make two separate products, one for the record and another for the stage. So although there is much studio craft at work here, things still move along two separate tracks. The production cycle varies from group to group, but it seems to go something like the Swans method described earlier: a simple demo is made, then turned into a full-fledged studio production. Even in the few years between Swans and Reznor, one sees a change which is indicative of a trend: while Swans would typically tour first and then record a song, Reznor and others now make the album first, then assemble the musicians to play a simplified live version.

The producers discussed in this chapter are creative powerhouses who have been successful in becoming not only a part of the chain, but really covering the whole process from idea to production to marketing through their own labels (Laswell's Axiom, Gira's Young God Records, Reznor's Nothing). As we have seen, their personal involvement in their productions tends to be high, although the similarity in sound between their own music and the bands they produce varies widely. In a conceptual sense, they are perhaps still transitional figures in the overall development from the back-seat producer of the old days to the contemporary composer–producer–performer of popular electronic music. This does not make their music any less important; indeed, most producers in the pop mainstream today are still well behind the advances in both concept and execution made by producers with the vision of a Laswell or a Reznor.

# 3   The Producer Takes Center Stage

## The Discothèque and *Musique Concrète*

Until this point, we have traced the development of the recording studio and the producer's role in it as though it were a tidy affair, a neat, logical progression which leads in a great arc from Fred Gaisberg with his phonographs to Trent Reznor and beyond. This is of course untrue. Partly because of the limitations of conventional writing, which is a monophonic medium, and also because of considerations of continuity, we have omitted the discussion of a phenomenon which began in the fifties and has grown exponentially ever since the advent of disco. That phenomenon is something which, for lack of a better term, we'll call studio music, that is to say, music which is meant to live primarily in recorded form; or, in more recent times, is performed by putting the studio on stage and mixing tracks in real time.

The antecedents of this music go back to France on two levels. On the level of "serious" music, there were the *musique concrète* experiments of French composers Pierre Schaeffer and Pierre Henry, who first composed music in which the studio was the performer. More precisely, they created music which was meant to exist exclusively in recorded form, as tape music. By freeing themselves of the notion of performance, they were able to manipulate not only the sounds of musical instruments, but of the real world as well. Their work was echoed in the United States by early composers of taped electronic music such as Vladimir Ussachevsky and Milton Babbitt.

At the level of popular music, France was where the replacement of live performers at public events by recordings began. Wartime Paris was the birthplace of the discothèque, a combination of the words "disque"

(record) and "bibliothèque" (library). According to Erikka Haa, author of
*Boogie Nights: The Disco Age*:

At the first discothèque in Paris patrons could order their favorite jazz records along
with their drinks. Dancing to recorded music was a form of defiance against the German occupation during World War II. The German forces outlawed dancing to
"American-type" music, which was considered decadent. With aid from the French
Resistance, many of these clubs went underground to escape the tyranny, and continued to provide the emotional release the citizens of war-torn France desperately
needed. After the war, these clubs continued to prosper. (Haa, 46)

What was born out of necessity gradually developed into a global trend: by
1980, there were over one hundred thousand discos operating worldwide
(ibid., 48). The music played at these events underwent major transforma-
tions over time. What had begun as a gathering place to listen to jazz
under the noses of the occupying Nazis eventually evolved into a mass
phenomenon. But although the stage was set in France, it is hard to make
an argument for a direct lineage.

**Disco: The "Producer's Genre"**

The musical genre known as disco gradually evolved out of soul (James
Brown), funk, and R&B, and is essentially an African-American contribution to music. Disco has been called "the producer's genre," because most
disco acts never performed on stage, but were instead the creations of producers, engineers, and studio musicians. This is not to say that disco was
entirely a studio medium: chart-toppers of the seventies such as the Bee
Gees and K.C. and the Sunshine Band did perform live. But the majority of
disco hits were studio productions which were presented to the public in
discos by a relative newcomer to the music business: the club DJ.

Why was disco almost exclusively about recorded music? One answer is
given by the Tom Smucker, author of the "Disco" article in *The Rolling Stone
History of Rock and Roll*:

According to legend, early disco took shape on Fire Island and in Manhattan at
places like the Loft and the 10th floor—part private clubs, part personal dance parties and part avant-garde proving grounds—because gay men couldn't get live acts
to perform for them. (Smucker, 562)

This may not be the only reason, though. The answer is also technological:
disco came along in the early seventies, which roughly coincides with the
advent of the twenty-four-track studio, the commercialization of synthe-

sizers, and the beginnings of console automation. It was therefore possible to make records in the studio which couldn't be performed live, as well as to remix and alter tracks.

Another reason for the ascendancy of recordings over live performance may have been the increasing importance of the DJ. Erika Haa writes:

The record hop of the 1950s carried his own 45-rpm records and provided the musical entertainment for banquets and weddings, mostly. . . . It was during the late Sixties that the DJ emerged from being an anonymous figure hidden behind a secluded booth (much like a movie-theater projectionist) to being a celebrity. . . . Discotheques were known primarily for featuring recorded music. The deejay became an essential element of discos, as the jukebox had been an essential element of juke joints previously. (Haa, 47)

What has made the DJ important is his ability to chain together record after record, building a large-scale ebb and flow which can (and does) go on for hours. In this sense he performs the function of an arranger and composer, structuring the flow not of any individual piece, but of a whole night. The uninterrupted flow of music is made more homogenous by the steadiness of the ubiquitous quarter-note pulse of the kick drum.

The ability of a good DJ to draw people onto the dance floor is based partly on how effectively he can set the pace for the audience. This audience, in turn, is in a sense the main attraction, because it is on stage (the dance floor). By using recordings rather than live performers, the focus is on the audience as the visual center of the spectacle. Use of recordings also means flawless performances, an immense variety of sounds, and top-quality production values. Also, as discussed below, the ability to re-edit and remix has become crucial.

### "I Feel Love"

A prime example of disco at its peak is producers Giorgo Moroder and Pete Bellote's "I Feel Love," with vocals written and performed by Donna Summer. Recorded in 1976, the song was an instant hit. The reputations of Summer and her producers had already been established the year before with the phenomenally successful "Love to Love You Baby," a song which featured "22 orgasms," according to *Time* magazine; Summer sold millions of records for a few years and made the cover of *Newsweek* in 1979, with the title "Disco Takes Over" (Haden-Guest, xxiv).

"I Feel Love" is a studio construct from beginning to end. There are no acoustic instruments other than Summers's voice on the track. The song whisks the listener off into its world as soon as it begins: the shifting kaleidoscope of synthesizers which make up the track's instrumentation enter full force over a full "four-on-the-floor" kick drum. Skillfully programmed analog synthesizers and various delays outlining a massive C–G–Bb ostinato shift subtly across the stereo field as various other oscillators chatter clavelike rhythms which never quite seem to repeat exactly. The top voice, an Eb which had indicated C minor, suddenly and quite effectively shifts up by half-step. The song shifts modes a couple of times, making simple yet effective use of major–minor mixture. Long, held notes shift timbre subtly, and syncopated synth voices fade in and out of the mix almost imperceptibly, providing the kind of subtle variety which is the hallmark of good minimalism. The overall effect is one of almost instant trance, and by the time the voice enters the listener is already hypnotized.

The studio treatment of the voice is almost diabolically skillful, and only becomes evident on close listening. Moroder and his team, notably engineer Harold Faltermeyer, managed to pull off a stunt most people only dream of, the sort of thing that seems like a great idea when someone suggests it but rarely actually sounds good when attempted. It's this: on every phrase, throughout the verse, Summer's voice starts out bone-dry. As the vocal line descends, it is first given a little chorusing, then a short reverb; by the time it hits the fourth bar, it's drenched in ambience. The effect on the listener is to perceive an added dimension to the performance without necessarily knowing why; it's a subtle and effective way of using virtual space to help shape a vocal line.

The no-nonsense construction of the song doesn't hurt, either. It's as simple as the layering of the synths which make up its sound is complex. The bare-bones three-note riff which makes up the sequence heard in the verse rises by step every eight bars until it falls back to the tonic in the chorus, which then restates the sequence's up-by-step motion twice, at a faster rate. The entire song consists of intro–verse–chorus–break–verse–chorus–out, its elegant symmetry clocking in at a surprisingly brief 3:22.

Moroder has a keen ear for detail, a superb sense of pacing, and a pop producer's ability to cut concise, compelling tracks. It seems that the way this music was made was that Moroder and company would come up with a complete track and present it to Donna Summer, who would then invent her vocal on top of it in the studio. Talking about how she came up with her vocal tracks in general, and specifically about "Love to Love You Baby," Summer says:

I would allow myself to feel the music, and be part of the music, until I felt the words. And I came up with whatever the song was about. And I never got that far with this song! I just got to the part where I kind of imagined myself as the character doing that . . . and not knowing that it was finished, not *thinking* that it was . . . thinking that I was going to go back and write the lyrics. . . . (Quoted in Haden-Guest, xxii)

Here again we have music which was not made to be performed live, and which doesn't even attempt to create the illusion of a "real" performance. Freed from the precepts of traditional sonic realism by predecessors such as George Martin, Moroder and other disco producers such as Nile Rodgers present the listener with musical artifacts which are forthright studio productions, and exist apart from the world of conventional live performance.

## Michael Jackson's Work with Quincy Jones

Michael Jackson's *Thriller* (1982) has sold over fifty million copies worldwide, and is the best-selling album in history (*Guinness Book of World Records*, 2003). For this reason alone, taking a closer look at Jackson's work with producer Quincy Jones seems worthwhile.

*Thriller* marked the high point of Jackson's career, a remarkable journey that began when, at age five, he began singing with his talented brothers as the Jackson 5. Originally from Gary, Indiana, the Jackson 5 were discovered and signed to Motown by its founder Berry Gordy in 1968. By 1970, they were topping the Billboard's pop charts with hits such as "ABC" and "I'll Be There." Described as a "miniature James Brown," Michael was one of America's biggest child stars, pirouetting, doing splits, and "able to hit any note."

In 1976, the Jackson 5 left Motown and as a result were sued over ownership of their name. This made it necessary to drop the "5" and become simply "the Jacksons." The group then worked with noted disco producers Gamble and Huff, famous for the "Philadelphia Sound" emanating from Sigma Studios. *Enjoy Yourself* (1976) was somewhat successful, even though it sounds generic. By 1978, the Jacksons were self-producing, and released *Destiny*. Sales of this album were not very strong. Around this time, Michael landed the part of the Scarecrow in *The Wiz*, a remake of *The Wizard of Oz* with an all-African-American cast. Quincy Jones was music director, and a rapport developed between the two men that led to collaboration in the studio. In 1979, *Off the Wall* was produced by Jones and released on CBS. It proved to be a smash hit, selling over nineteen million albums worldwide. Overnight, a young adult Jackson became a pop sensation, transcending his already remarkable career as a child star.

*Off the Wall's* first track, "Don't Stop 'Til You Get Enough," is a dance classic. Coproduced by Jackson, it has both feet on the disco floor. The music is very 1979, and shows the influence of the Gamble and Huff approach: pristine sounds, carefully delineated in the mix, performed to perfection. The instrumentation is vintage late-seventies R&B/disco:

drums

percussion

electric bass

clean, funky (Strat) guitars panned left and right

electric piano

horns

strings

The form of the song also evokes Motown's Holland–Dozier–Holland writing and production team, known for "rarely using standard song forms, opting instead for a simpler, more direct *ababcc* pattern, anchored by an endless refrain of the song's hook line" (Curtis and Anthony, 280). "Don't Stop 'Til You Get Enough" "provided the soundtrack for every discotheque as the Seventies melted into the Eighties" (ibid., 651).

The year after Michael's *Off the Wall*, the Jacksons released the self-produced *Victory* (1980). Again, it enjoyed only moderate success. Michael's star now far outshone the other Jacksons, who were feeling increasingly irrelevant.

The recording of *Thriller* began in April of 1982. Michael Jackson contributed three songs, which would all become smash hits: "Billie Jean," "Beat It," and "Wanna Be Startin' Somethin'." One can hardly overstate the influence of this album. While it wasn't truly revolutionary (producer Quincy Jones describes himself as a "culminator"), it successfully melded a number of different styles, and set the course for dance pop for the next two decades. *Thriller* transcended disco and moved into a new, hybrid realm that successfully blended R&B, soul, funk, and even heavy metal.

The cast credited on the album lists over forty musicians, most notably four members of the then hugely popular band Toto ("Africa"), Paul McCartney, and songwriter Rod Temperton, who wrote the title song.

Jones carefully calculated his commercial approach:

I think half the producer's job is assuring you've got the right songs. And that's the biggest conflict we have sometimes with the artists, because they're in love with everything they write, and it's hard for them to kill their babies, you know. But at some point you have to choose a group of songs and say, "this is it."

I think for a record to penetrate like this record did, you have to go for the throat in four, five, or six different areas . . . You do a rock-n-roll kind of a trip, you know, you do your "AC" thing, and dance stuff, R&B, soul, whatever you want to call it.

But he is quick to say that the outcome is ultimately beyond his control:

Anybody who tells me they know how to make an album like [Thriller] is just lying, because you can't plan it. Nobody knows how to aim at anything like that. You can't, man, it's got something to do with divinity. (Transcribed from recording, Thriller Anniversary Edition, interview with Quincy Jones)

### "Billie Jean"

A revealing home demo of this song is included on the 2001 re-release of Thriller. Made in Michael Jackson's home studio, the demo shows what Jackson brought to the big studio (L.A.'s Westlake). Like demos of other now-famous songs ("Strawberry Fields," for example), this tape sketch shows that Jackson is composing in his studio. This is not surprising, given that he had previously made fleshed-out demos for Off the Wall (Off the Wall Special Edition, bonus tracks: demos of "Don't Stop" and "Workin' Day and Night"), and coproduced the Jacksons' Victory in 1980.

The demo's simple drum beat, which sounds machine-made (probably on a Linn Drum), is carried over to the eventual track, played by N'dugu Chancler. The sounds on the final track are real drums, played with commitment by a master drummer: they sound confident, hard, and precise. They are also tuned higher and have a brighter sound than the drums heard on the demo. In the third bar of the album version, the kit is joined by a shaker that duplicates the hi-hat's steady eighth-note pattern. The careful attention to both the sound of the percussion and its feel is reminiscent of early sixties production. It's a reminder that Quincy Jones draws on a vast amount of studio experience. As head of Mercury Records A&R, he produced Lesley Gore's 1963 hit "It's My Party," a song which is from the same era as Phil Spector's girl-group hits (see the discussion of "Be My Baby," in chap. 1, in which the percussion is described in detail).

The motoric synth bass line, which sounds indebted to the German tradition of analog sequencing (Kraftwerk, Giorgio Moroder), is also carried over almost literally from demo to final product, but with important differences in the details of the sound. The groove of "Billie Jean," while a steady stream of eighth-notes, is soon propelled by synth stabs on the "one" and "two and." In the demo, these accents are ignored by the bass; subtly but persuasively, though, the final version has the bass synth emitting a slightly menacing, yet warm growl on these same accents. This serves to set up the space for the chordal synths in subliminal fashion. Additionally, there is a shift from demo to final in the main bass frequency from 140 to 90 Hz (visible when analyzed by computer), which causes the synth to sit better in the mix.

As artists and producers well know, it can be very hard to keep the feel of a demo once the tracking and the pressure begin in earnest in the big studio. Jones must be aware of this syndrome, for he chooses to keep certain details of the demo which one might have expected to be deleted in a polished production. It's as though these little touches, some of which were quite possibly accidental, are somehow necessary to the overall picture. They are replicated with great care. For example, Jackson's impromptu-sounding "sshh" (with added delay) in bar seven of the demo (0:12) appears in a tighter, better-recorded fashion in almost exactly the same spot on the final "Billie Jean" (it's moved a little before the beat, in keeping with funk practice). A pair of clipped, off-beat sighs herald the vocal entrance. Right from the start, Michael Jackson's delivery is tight, emotionally charged, and utterly in the groove. Surprisingly, though, his voice is *in* the track, not in front of it. As the song progresses, the listener is enveloped by the finely woven tapestry of Jackson's multitracked voice. The performance and production of the vocals blend seamlessly to create a unique sonic signature. Throughout the track, there is a subtle interplay between solo voice; a dry, foregrounded chorus at the end of most phrases; a background chorus; a main chorus; and assorted whoops, yelps, and shouts that add up to a percussion section of their own. Where the demo shows Jackson trying out ideas and searching for his vocal, the album features a thoroughly polished arrangement of unique character.

The synth performance is tightened up a little on the final version, but it's again very close to the demo—it's the same chords and the same sound, with just a little less decay on each chord. At the end of the verse, the final version is enhanced by repeating the synth chord to fit the groove; there's also now a subtle and quite effective little dynamic swell. It marks the end of the verse, and recurs faithfully at each juncture throughout the song.

Viewed individually, these details may not seem particularly significant. Yet it's telling how every single change is clearly an improvement, and serves to strengthen the song overall. Everything in the album version is carefully sculpted: every single sound is clear, and there for a reason. Yet although there's not a hair out of place, the overall impression of the track is that it is organic.

At the bridge (1:09), an electric piano plays a contrapuntal line over the bass, and is echoed by a brassy synth turn. One hardly notices that the main synth, which has been faithfully holding down C#–D#–E throughout, is now absent. The bass line has changed, but it's almost imperceptible; it will return to its looping cycle once the chorus enters. As we build to the chorus, the drummer contributes splashes on a China-type cymbal on the four of every other bar, and opens the hat just long enough for a hit on the four "and"; the hi-hat intensifies its syncopations as eighth-notes on the floor tom lead up to the climax. A calliope-like synth in the background adds to the tension with a dominant seventh chord, its upper voices shifting up an octave as the tension rises. The vocals, which began the bridge as Michael Jackson solo, gradually build into frenzied shrieks and moans before giving way to the payoff.

The chorus is thoroughly satisfying, taking the song to an ecstatic plateau as Jackson delivers his message: "Billie Jean is not my lover / She's just a girl who claims I am the one / But the kid is not my son / She says I am the one / But the kid is not my son." The vocals are massed thirds and sixths mixed with slinky unisons. The Em–F#m–G–F#m synth mantra is back; the chords are now held longer, filling in the previously open space. A funky guitar, heard on the demo in exactly the same fashion, plays a clipped, repeating figure rich in overtones. It's very clean and a bit edgy, in the manner of a Stratocaster.

The second verse features a new synth line, again carried over from the demo. By now it's obvious that the demo has served not only to outline the song idea, but also as a template for the instrumentation of the final track. The nearly exact timbres of the bass, electric piano, and various synthesizers have been copied and refined on the album. Jackson has effectively coproduced the track by providing the session musicians with not only parts, but sounds as well. He has composed the track in his project studio, and the big studio is a place of instrumental fine-tuning; most of the work at Westlake Studios seems to have been done on the vocal arrangement, which is quite elaborate and virtuosic, as we have seen above.

At the second bridge (2:31), a full string section suddenly appears, taking the song's energy up another notch. Looking back on the first bridge, it's now clear that the space had been left open for just such a thing. This has happened before, with the synth chords: once the space is filled in, it's as though it was always planned that way, but the producers hadn't tipped their hand until the time was right.

The strings sound quite lush, and are reminiscent of a big-budget Hollywood production. The arrangement is slick, and like everything in this track, it works. It's unusual to find such a combination of elaborate synthesizer work and real strings. For one thing, it's quite expensive. String sections remain a major production expense today, but the cost of using synthesizers has dropped dramatically. In 1982, a standard Prophet 5 polyphonic synthesizer cost about $4,500 (equal to about $12,000 today), and professional synthesists amassed rooms full of synths made by companies such as Oberheim, ARP, and Moog. On this project, the producers had the luxury of an unlimited sonic palette. Quincy Jones and his crew have come up with a beautiful combination of polished string arrangement and exciting synthesizer work. The blend is perfect, and sounds contemporary twenty years after it was recorded; at the time, like everything on this album, it was state-of-the-art.

The second (and final) chorus begins at 2:48. It lasts over two minutes, making up about forty percent of the total track time of 4:53. As the song continues to intensify, the synths and strings trade licks, or share phrases. At 3:24, for example, a brassy synth plays an expressive ascending lick for two beats, and is immediately answered by a descending string figure. The synth says "Stevie Wonder," while the strings say "James Bond movie as scored by John Barry." As Albin Zak has pointed out (Zak, 81), pop music deals in tropes which have sonic signatures—it's not just about the notes played, it's about the way they were recorded and the resulting distinctive timbre.

In the fifth minute of "Billie Jean," all of the elements previously brought into play interact. The guitar plays variations on its basic figure, but always stays within its assigned role as a rhythm (as opposed to solo) instrument. The strings and the synths go from trading licks to blending into a convincing unit. Jackson's vocals improvise freely and on multiple tracks over the lush arrangement. The overall impression is of a texture which, although derived from disco, soul, and rock, has a distinctive character. It's not overly glossy, as Gamble and Huff could be. This is more than ear candy—Jackson and his players project musical commitment and emotional depth in the context of a pop song. And listeners can sense this, whether or not they "know" anything about music.

Ultimately, what is striking about this production is that behind its apparent simplicity and transparency, there is a team of over forty musicians and technicians, guided by one of the most experienced and successful producers the music industry has known.

## Kraftwerk and Conny Plank

One of the first all-electronic pop tracks to be commercially released was a one-off by a group called Hot Butter. The song, entitled "Popcorn," was perceived as a novelty at the time. With her *Switched-On Bach* (1968), Wendy Carlos had shown that making all-electronic music that would be listened to by many was a viable proposition. But it was German groups such as Tangerine Dream and Kraftwerk that made an impact on popular music with all-electronic recordings. Where Tangerine Dream tended to translate orchestral textures into synthesized "cosmic music," Kraftwerk probed what they called "machine music" with increasing rigor, culminating in the practically single-handed invention of pure electronic dance music.

The duo behind Kraftwerk were Florian Schneider and Ralf Huetter, who originally met as students at the Düsseldorf conservatory. They started out as a rock outfit; their first group, Organisation, released an album called *Tone Float* in 1970 (Bussy, 26). Their producer was Conny Plank, an avid experimenter who was a radio sound technician by training. Plank was also associated with the experimental rock group Can, as well as Cluster and later DAF. D. Strauss writes that Plank "was a master of psychedelic effects, such as phased drums, and an inventor fascinated by technology, such as rhythm boxes" (Strauss, 62). Working with Plank, Kraftwerk honed an increasingly industrial, electronic image, gradually moving away from their early imitation-British sound (in Kraftwerk's early period, Huetter played flute, "sounding like an exaggerated Ian Anderson" [ibid.]). In time, Schneider and Huetter set up their own studio, KlingKlang, which was for their exclusive use. Their breakthrough album, *Autobahn* (1974), was made with Plank, as were *Kraftwerk 1* (1971) and *Kraftwerk 2* (1973). After *Auto-bahn*, Kraftwerk began to produce themselves on a permanent basis. They became known as denizens of the studio, mysterious entities who created their purely electronic music behind closed doors.

There was little distinction in Kraftwerk between the music and the studio. In effect, advances in recording technology would become the raison d'etre of the group's existence. Eventually, they became obsessed with producing music that almost sounded as if it had been created by machines—not just musicians who were also studio engineers, but more like sound engineers who happened to produce music. This led to the logical conclusion that the studio was a musical instrument or member of the group in its own right. As they would put it, "we play the studio." Ultimately, as the Kraftwerk sound developed, the studio became a kind of technical laboratory, claiming that they were not so much entertainers as scientists. (Bussy, 25)

Albums such as *Trans Europe Express* (1979), *Man-Machine* (1980), and *Computer World* (1982) followed. All of these works are carefully constructed studies in minimal electronic music with a pop sensibility. Kraftwerk has been very influential, pointing the way not only for electronic pop and techno, but also being sampled by early rap/electro artists such as Afrikaa Bambataa.

### Hip-Hop and the Rise of Sampling

In the Bronx and East Harlem of the mid-seventies, street culture spawned a genre of music with certain similarities to disco which came to be known as rap or hip-hop. In terms of the creation of the recorded artifact, this too is a producer's genre. Music in the conventional sense is relegated to a secondary role; in hip-hop, it's usually the vocals that distinguish one track from another, that make it unique. Musically, the requirements are deceptively simple: it just has to have a good beat and a strong hook. The tracks used need not be originals; hip-hop was the first form of popular music in which the wholesale appropriation of someone else's track became standard, if not always accepted, practice. Two of the earliest rap hits, the Sugarhill Gang's "Rapper's Delight" and Grandmaster Flash's "White Lines," simply copied the music of other groups and rapped over it. In the case of the Sugarhill Gang, the source material was Chic's "Good Times"; Grandmaster Flash helped himself to an entire track by Downtown funkers Liquid Liquid, and renamed it "White Lines," a move which eventually cost him his record royalties for that production in court.

The precursors of rap were disco, funk, and Jamaican dub. Disco was particularly close; "Rapper's Delight," with a disco track as its instrumental backing, was the single that brought rap into the mainstream, selling an estimated fourteen million copies. Disco had a different audience than hip-hop, which had existed as an underground form since about 1970. By 1979, the year "Rapper's Delight" appeared, disco had gone mainstream on one hand, while it was also associated with the jet-setting beautiful people (Studio 54, etc.) on the other. It did not represent the voice of the inner city, of working-class and disadvantaged America. Youths in poor neighborhoods, who generally did not have access to then-expensive studio time, started creating a new form of music at street parties.

The reason for the widespread use of previously existing tracks may be traced back to the origins of the form. In the late sixties, a young Jamaican named Clive Campbell, later known as Kool Herc, arrived in New York. By the early seventies, Kool Herc was DJing in the streets of the South Bronx, tapping into city lightpoles for electricity and spinning records. He had brought with him from Jamaica the style known as "toasting," which involved the DJ talking over the break section (drums and bass only) of a record. Kool Herc was one of the first to do this with the Shadows' "Apache," which has remained an often-sampled and covered break to this day.

In musical terms, sampling began with hip-hop, before samplers even existed. To get their two- or four-bar phrases to yield special effects such as looping, DJs invented the two-turntable setup, consisting of two turntables, a microphone, a mixer, and some cardboard or similar material (the cardboard is placed between the record and the turntable in order to permit the platter to revolve at a steady speed underneath the action). Grandmixer DST, who worked on "Rockit," explained some turntable techniques to writer Stephen Hager in 1984:

When he is cutting, DST repeats a drum beat or musical phrase by shifting the record back and forth while keeping the needle in its groove. Moving the record clockwise produces the normal sound heard when playing the record. Moving the record counterclockwise produces the strange swishing sound of the record being played backwards. By carefully controlling the speed of the turntable, DST can turn his records into miniature synthesizers or drum machines.

When he is backspinning, DST needs two copies of the same record. Unlike cutting, backspinning does not alter the sound of a record—it merely extends what's already on the record. For example, suppose DST wants to extend a brief drum beat several minutes longer than it appears on a record. He merely plays the beat on the first record while "backspinning" the second record to the right location. By the time the beat has reached the end, DST is ready to kick in with the second record. The process can continue indefinitely. Beginners often put arrows on their labels, indicating the approximate location they want to backspin to. More experienced DJs learn to read a record as easily as the face of a clock [ . . . ].

When he creates a mix, DST attempts to maintain a steady dance beat through several different records. Each record must be synchronized by adjusting the pitch control on his turntable. (Hager, 41)

While many hip-hop productions have replaced these seat-of-the-pants techniques with samplers and computers, scratching remains a rough-and-ready, highly effective way to make electronic music in real time. The U.K. certainly seems to like it: lately, more turntables have been sold there than electric guitars (Clark).

Partly because this was the technology that was available, and partly because it was just a cool thing to do, hip-hop introduced the recycling of previous recordings which is ubiquitous today. Afrika Bambataa recalls the early days:

I started playing all forms of music. Myself, I used to play the weirdest stuff at a party. Everybody just thought I was crazy. When everybody was going crazy I would throw on a commercial to cool them out—I'd throw on "The Pink Panther" theme for everybody who thought they was cool like the Pink Panther, and then I would play "Honky Tonk Woman" by the Rolling Stones and just keep that beat going. I'd play something from metal rock records like Grand Funk Railroad. "Inside Looking Out" is just the bass and drumming . . . rrrrmmmmm . . . and everybody starts freaking out.

I'd throw on *Sergeant Pepper's Lonely Hearts Club Band*—just that drum part. One two, three, BAM—and they'd be screaming and partying. I'd throw on the Monkees, "Mary Mary"—just the beat part where they'd go "Mary Mary, where are you going?"—and they'd start going crazy. I'd say, "you just danced to the Monkees." They'd say, "You liar. I didn't dance to no Monkees." I'd like to catch people who categorize records. . . . (Quoted in Longhurst, 152)

The producer is an integral part of the act of creating hip-hop tracks. He often creates the instrumental parts. Increasingly, "writing" has come to mean the deft combination of samples from various sources, and the skilled manipulation of technology such as samplers, synthesizers, mixing boards, and computers. As was often the case with the vocalist in disco, rappers usually perform their vocals over tracks put together (composed) by the producer.

By 1982, rap tracks such as "Planet Rock" by Afrika Bambataa were beginning to explore the juxtaposition of disparate elements over a unifying beat, a practice which electronica was to take to another level in the nineties. Borrowing from German electronic group Kraftwerk's *Trans Europe Express* and *Numbers*, and rapping on the same track, Bambataa captured the cross-fertilization taking place in New York clubs such as the Roxy, where both white New Wave kids and uptown black and Latino clubgoers hung out. The impact on the music was to make it more universal, and also more visible still to white mainstream audiences.

## Hip-Hop in the Late Eighties

The eighties also saw the rise of formidable producers such as Rick Rubin, who produced multiplatinum acts such as the Beastie Boys and Public Enemy. The Beastie Boys' "Fight for Your Right to Party" was the statement of the spoiled, trashy white kid image they projected; Public Enemy, known for their Black Nationalist stance, countered with "Party for Your Right to Fight" in 1988. Rubin had started out as a DJ while a student at New York University (Ogg). His love of both punk and rap inspired him to create records which crossed over between rock and rap. In 1986 he produced Run D.M.C.'s cover of white Boston rockers Aerosmith's "Walk This Way," which sold millions of copies and further enhanced rap's mass appeal.

### "Bring the Noise"

Public Enemy's *It Takes a Nation of Millions to Hold Us Back* raised the bar for hip-hop production. Producers Hank Shocklee and Carl Ryder, together with Public Enemy's conceptual mastermind Chuck D, used their extensive knowledge of records and their skill with turntables and samplers to craft a sound for the album that was altogether new. Rick Rubin was executive producer for the project, which means that he oversaw the production as a whole but did not get involved in the actual creation of the music.

The situation with Public Enemy was unusual in that it was a band of DJs and rappers, all of whom had input into the creation of the album's sound. It's a sound which is dense and more complex than most hip-hop. Without listening closely, it's easy to mistake it for noise, but in reality it is a highly structured, cohesive style. It's a sonic *tour de force*, especially for the tech-nology available at the time. Such density falls into a particularly African-American tradition. Musicologist Olly Wilson writes:

The drum set of early twentieth-century Afro-American music maintains indepen-dence by means of timbral differentiation. The sound ideal of the West African sphere of influence is a heterogenous one.

Another common characteristic is the high density of musical events within a rel-atively short musical space. There tends to be a profusion of musical activities going on simultaneously as if an attempt is being made to fill up every available area of musical space. (O. Wilson, 15–16)

The second track of the album, "Bring the Noise," opens with a sample, probably of Black Nationalist leader Louis Farrakhan, repeating "too black . . . too strong." A hard-hitting four-bar intro follows, establishing the sound for the whole set: rapper Terminator X riffs over an arresting combination of horn and drum samples, which have been cut up and recombined to create a potent new groove. Underneath it all an almost subliminal bass is churning away, pumping out a repeating D–C–G–A riff.

Chuck D appears as the verse kicks in, his accomplished rhymes enhanc-ing the sure-footed quality of the production overall. Yet the style of the rap is comparatively light compared to later West Coast groups such as N.W.A. While the words convey a strong political message, their delivery is more like "Rapper's Delight" than the harder-edged vocal style that came into vogue in the nineties with producers such as Suge Knight and Dr. Dre.

The vocal is just barely in front of the backing track, which features a syncopated interplay between drum and horn samples. The complex cross-rhythms branch out in all directions: the instruments are but one layer, with Chuck D adding another dimension to the rhythmic feel with his vocal. On top of all this, samples of undeterminable origin add strange, unpleasant noises to the mix which enhance the unsettling feel of the track as a whole.

Sixteen bars later (0:49), the rap breaks into the chorus, a feature unusual for rap at the time. For all the surface chaos, there's a tight, rather conven-tional structure underneath which holds "Bring the Noise" together. This is an aspect of the production which probably influenced artists such as Trent Reznor. However, this is rap, and there is no melody and very little in the way of harmony in the conventional sense; the music is based on repeat-ing beats, simple riffs, and the vocal.

The chorus provides an effective contrast, while still developing the material presented in the verse. The mixture of scratchy horn samples and drums continues, now with an added element of some parts being played backwards (literally backwards, i.e., samples with their volume envelopes reversed) while others stay normal—a sort of time-warp drum set is heard. Chaotic scratching appears from time to time, and before we know it we're catapulted into a quotation from James Brown, the beat sampled from "The Funky Drummer." The famous groove is made more urgent and intense by the addition of a synthesized kick drum playing continuous 16th-notes for the duration of the bridge (1:07–1:16). After repeats of verse and chorus with different lyrics, an instrumental verse happens at 2:43. The instrument is undefinable, but it sounds like a snippet of a reed instrument grabbed off of an old record and performed as a scratch; on every four, a crowd roars approval of the performance (also probably a sample). All the while Chuck D and Terminator X are holding forth about the struggles involved with creating their music and making their mark in a world dominated by rock: "You call 'em demos, but we ride limos, too / Whatcha gonna do? Rap is not afraid of you . . . Run D.M.C. first said a deejay could be a band."

*Nation of Millions* was groundbreaking in that it took the content of rap to a new level of meaning, but also because it was one of the first top-selling pop albums to be made up entirely of samples and drum-machine beats. Where earlier acts such as Sugarhill Gang, Grandmaster Flash, and Run D.M.C. had relied on live musicians playing conventional instruments such as guitar and synthesizers for at least part of their sound, on this album Public Enemy relies entirely on manipulating samples. Their ability to transform the sound electronically was technically limited compared to the technology which became available in the recording studio of the nineties. This hardly comes across as a limitation in their work, though: Public Enemy's collective imagination more than overcame any limitations imposed by technology.

## The Hip-Hop Producer Today

Hip-hop evolved into a global phenomenon during the nineties. From its humble beginnings as Bronx street-party music, the form evolved into a billion-dollar industry, reaching beyond music into fashion and films. The hip-hop producer is visible beyond hip-hop. His position is widely recognized in the field, as well as in the media. In September of 2000, ABC aired a program called "Russell Simmons' Countdown of the Top Ten Producers of Hip-Hop," which featured excerpts from the work of top rap producers, as well as interviews with them. Although names such as the Trackmasters and Swizz Beats may not be household names, some of the artists they've produced are: Whitney Houston, Mariah Carey, Aretha Franklin, and other top stars of the pop firmament have all worked with hip-hop producers. Top producers Timbaland, Manny Fresh, and DJ Premiere have each sold more than ten million records.

What is most interesting about the current scene is that producers should be treated as stars in their own right, with one after the other being interviewed on *Countdown* about how they put their work together. The sense one gets is that everyone is now using computers running sequencing and recording software, and that the term "track" in hip-hop denotes the music bed, which is created before the vocal. The Trackmasters relate that their standard procedure is to meet with a new artist, and then disappear into the studio for two weeks or so, "work twenty-hour days, sleep four or five hours, then get back to work," and come up with a backing arrangement to present to the artist. This method of working is reminiscent of how Giorgio Moroder and other disco producers went about making music in the late seventies; at the time it was unusual, but now it's the way most pop music is created.

Hip-hop is here to stay; it's the main force in contemporary popular music, and the producer is in the driver's seat. At the time of this writing, the number one record on the pop charts is Outkast's *The Love Below/Speakerboxx*, a double album by Andre' 3000 and Antwon Patton. What is striking about this mainstream album is that it is produced by the artists themselves. In a CNN interview (Feb. 6, 2004), Andre' 3000 says:

If you have an idea in your head and you already know what you want it to sound like, it doesn't make sense to me to go to someone else and say "OK, this is what I want it to sound like."

What makes a statement like this possible is what has gone before it. One thing that comes to mind is the image of Michael Jackson composing his material in his home studio in 1981: he's not writing on paper, he's committing synthesizer sounds and drum-machine beats to tape. There's also Brian Eno and Trent Reznor's way of working, i.e. with pop artists composing in the studio using electronic sound sources (synthesizers, drum machines, samplers, computers). Here we again have the producer as composer in the truest sense.

## Electronica

As hip-hop developed in the U.S., England was watching and listening, and soon began to develop its own brand of popular electronic music. In the late eighties and early nineties, the first techno records started to appear, and the rave became a popular dance event all over Europe. Experimenters such as Goldie and Roni Size started making records which fused the energy of funk and hip-hop with an experimental sensibility, speeded everything up to 150 bpm and above, and called it "drum'n'bass." Soon a myriad of substyles such as trance and ambient (the latter being originally created by Brian Eno in the mid-seventies) were springing up all over. After a while journalists started using the term "electronica" to describe the movement as a whole, an umbrella term which encompasses all the various subgenres of popular electronic music.

The main difference between electronica and hip-hop is that the former is essentially instrumental music, and its creators are often anonymous. In some cases, they are not even visible on stage during their performances, as was the case with the French producer–writer duo Daft Punk when they performed at the Hammerstein Ballroom in New York in 1998.

The cult of personality and boasting which is part of hip-hop culture is anathema to electronica, where there are hardly ever any vocals, except perhaps for an emblematic phrase here and there chopped up and often treated in some way for musical effect. Yet to a large extent, electronica grew out of hip-hop: the primacy of the breakbeat is evident in major acts such as Prodigy and the Chemical Brothers, as is the occurrence of samples taken from previously released recordings.

The idea of making sample-based records was taken further by electronica, which legitimized sampling as an art form by adding the element of transformation. Where the hip-hop track still relies heavily on beats and scratching, the best electronica artists such as the Orb combine these elements with highly sophisticated electronic treatment of the source material, as well as the timbral colors of vintage analog synthesizers such as the ARP 2600. Some of the best electronica artists put their samples through so many different electronic processes that the original sound is hardly recognizable. Although the use of techniques such as resampling, filtering, shuffling, and the like partly developed as a way to keep from getting sued for copyright infringement, the results have propelled the development of this kind of electronic music into interesting places. Creative duos of producer–performers such as England's Autechre and the Chemical Brothers or France's Daft Punk and A.I.R. are the apotheosis of producer-made music. All of these bands are acts which create their music in the studio without the aid of a singer. They are also live performers who bring their project studios onto the stage with them and tour internationally, performing to packed houses around the world.

One of the best-known and most successful of these groups is London's Chemical Brothers, consisting of writer–producer–engineer–performers Ed Simons and Tom Rowlands. The two met while studying history in Manchester in 1989, and soon began DJing at parties together. By 1994, they were at the forefront of the English dance music scene, and had started their own subgenre of electronic music, which critics took to calling "Big Beat." According to one writer, they were "joining the dots between acid house, hip hop and rock" (Raft, 1). Simon Reynolds writes that

The Chemical Brothers stay close to the most radical aspects of house and techno; they mostly shun songs and vocals, and rarely resort to melody, yet still manage to enthrall though texture, noise, and sheer groove-power alone. (Reynolds, 383)

Their DJ sets at London's Heavenly Social, a well-known dance club, were frequented by many of the U.K.'s top pop stars. Tricky, Primal Scream, and Paul Weller were among those who showed up to take in the scene, and especially the sounds (ibid., 3).

Along with their activity as DJs, which was giving them exposure on the hip London scene, Simons and Rowlands were toiling away in their project studio. After releasing a string of successful singles and doing remix work for other artists, they released the first Chemical Brothers album, *Exit Planet Dust*, in 1995. It was an immediate success in England, and went on to sell over a million copies worldwide. They have released two albums since, both to critical and public acclaim: *Dig Your Own Hole* (1997) and *Surrender* (1999).

*Exit Planet Dust* was the album that put the extraordinary production duo on the musical map. At the time, not very many electronic groups had succeeded in sustaining a full album's worth of music, let alone in making a continuous, compelling musical experience last so long (49:26). Another unusual feature of this music was that although it made great dance music, it could also be listened to on its own. This gave rise to the term "I.D.M.," short for "intelligent dance music." (The term is most often associated with England's Warp label, which features artists such as Squarepusher and Autechre. The Chemical Brothers were signed to Virgil Records in 1993.)

The album's segues from track to track are ingenious, and reminiscent of the sort of "glue" Giorgio Moroder came up with for Donna Summer on *On the Radio: Greatest Hits Volumes I and II* in 1979. It's hard to tell where one track ends and the next one begins, the last sound of one cut often being the first of the next one as well. Tempos are meticulously matched; if there is a tempo change, it feels natural.

But there the similarities end. The Chemical Brothers are mostly funk-based, with excursions into the sort of latter-day psychedelia associated with trance and acid house. Although they are said to be part of a sound known as "Big Beat," their style is really all their own, a potent brew of sounds that is fit to induce excitement and hypnosis at the same time. The textures the Chemical Brothers produce are elusive. A first attempt at close listening to a track such as "Leave Home" (the first cut on *Exit Planet Dust*) can have the effect of deflating the music, because it seems too simple to work. Everything happens strictly in groups of four. New parts enter predictably on the first "one" of any given four-bar group. The pulse is entirely regular, and is always present at some level.

The reason "Leave Home" works so well can be explained in part by a statement made by another contemporary English duo of electronic composer–producers, Autechre. In a 1995 interview, Autechre member Sean Booth said: "We find that things that are sort of perfect but fucked up are pretty cool" (quoted in T. Palmer). The Chemical Brothers seem to espouse this same philosophy. Within the rigidly defined time of their sequences, there is an abundance of grit and randomness. But the chaos is captured and repeated, and thus cast in an entirely different light. The development of studio technology since 1970 plays an important part in making such an aesthetic possible. If one has, say, an ARP 2600 synthesizer, which was manufactured from 1971 to 1980, one owns a machine which is capable of almost infinite sonic variety. Eno's embrace of the vicissitudes of his slowly failing vintage 1971 VCS-3 is also well known.

These machines make great sounds, but it's almost impossible to get them to do the same thing twice. In the days before affordable samplers, using synthesizers was tricky. I used to own both a Korg MS-10 and an ARP

2600, and I well remember how frustrating working with them could be. The process had a poignancy about it; the best one could do was capture the sound on tape, and then the first-generation original was gone forever. The unrepeatability in real time of complex patches was due to the drift inherent in analog equipment, made worse by aging. But with the advent of samplers in the two- to three-thousand-dollar range, which happened around 1986 with the introduction of the Akai S900 sampler and others, it became possible for just about anyone who was serious about their music to work with samples of anything. As we have seen, hip-hop groups such as Public Enemy broke new ground with their practically exclusive use of samples over an entire album. While their samples tended to be of records, not of synthesizers, the principle was the same. These developments did not go unnoticed on the budding English electronica scene.

A look at the Chemical Brothers' equipment list, circa 1998 (Raft, 4), bears out the mix of analog and digital:

*Computers*

Apple Mac G3 266MHz

Apple Mac LC475

*Software*

Steinberg Cubase VST 3.52

Steinberg Recycle

Steinberg Rebirth

*Sequencers*

ARP 1603 analog sequencer

Doepfer maq 16:3 sequencer

Electro Harmonix analog sequencer

Akai MPC 3000/drum machine (16Mb)

*Samplers*

2 × Akai 3200XL

Akai S1000 (fully expanded)

Akai S2800 (6Mb)

Akai X7000

Emu E64

*Synthesizers*

ARP 2600

EMS synthi

Octave CAT

Octave Kitten

Korg mono-poly

Korg MS10

Roland SH101

Roland Juno 106

Electro Harmonix minisynthesizer

*Guitars*

1963 Fender Jazzmaster

1971 Fender Telecaster

*Processors/effects*

Sherman filterbank

Electro Harmonix bass microsynthesizer

Electro Harmonix guitar microsynthesizer

Electro Harmonix electric mistress

Electro Harmonix tone bender

Electro Harmonix space drum

Ibanez analog delay

Schaller tremolo

Boss heavy metal pedal

Morley wah/Dist/Boss Heavy metal

Alesis Quadraverb

*Recording*

Tascam DA88 digital multitracker

Mackie 1604 mixer

Kawai K4

This setup is rather typical of the modern studio geared toward the production of electronic dance music. In essence, the computer controls everything else, with the exception of analog equipment such as the synthesizers and the mixing board. Once everything has been sampled and sequenced, it is recorded to the multitrack digital tape deck (DA-88). The live parts (guitar, bass, occasionally vocals) are usually added onto the tape as well, unless flexibility in the arrangement process is important. Going straight to tape is simpler than tracking direct to hard disk; the tape deck is more reliable than the computer, which can be tricky. But obsolescence is also a big problem. Is the tape format more likely to be around in ten years than a version of Cubase, or the samplers which can still play the sequences exactly as they were made?

"Leave Home" is perfectly quantized and orderly on one level, yet full of bizarre noises of uncertain provenance and makeup on another. Rowlands and Simons combine this recycling of vintage synthesizer sounds with the skilled manipulation of drum samples, as well as actual guitar and bass guitar playing.

On another level, the Chemical Brothers use subtle arrangement techniques to move the mix along. At first, the opening section of "Leave Home" seems simply to progress by addition. But riffs and sounds are being transformed as the track progresses. After the first eight bars, the synth drops out, and is replaced by a funk guitar riff. Four bars later, the guitar is joined by a bass playing a 16th-note ostinato. When the drum beat enters, the riff is simplified, and is made to fit in with the busier whole.

Another compositional device the band is fond of is the buildup of tension by use of a rising sound coupled with intensifying drums, especially electronic snare (in fact, electronica generally eschews anything but kick, snare, and hi-hat; this is especially true of drum'n'bass, where tom-toms are almost never heard). In "Leave Home," this happens at 2:13 and 3:51, lasting eight bars each time. The effect is to practically shoot the return of the verse (the obsessive E-flat bass riff) out of a cannon each time it returns.

Elsewhere, the arrangement is subtly varied. At 0:54, for example, the beat continues as before, but a second snare is added to intensify the groove; as the track progresses, subliminal synth noises enter and disappear. Variation is always introduced at beginning of a four-bar phrase, and lasts either four or eight bars. Again, the tension between change and stasis, order and chaos is established and explored by means of strange noises being assigned exact places. In the last few years, electronica acts such as Squarepusher, Autechre, the Chemical Brothers, and Daft Punk have been performing with their entire project studios on stage. This has given rise to the category of the "performing engineer," as an April 1999 cover of *EQ* magazine announced. "Performing producer" might have been more accurate, or perhaps some term which could somehow capture the fact that composer, producer, engineer, and performer have all been melded into one function, usually performed by two people. The groups go through an elaborate process of preparation, and then present their results in performance as well as on disc.

From the beginning, popular electronic music has stressed the element of performance, as exemplified by the record-scratching developed in the Bronx of the late seventies. Yet it may seem strange that anyone would bother to deal with the elaborate setups necessary to achieve a live project studio on stage. After all, for the most part all that's really happening is that sequences are being played back. But attending, say, a Daft Punk show proves to be enlightening in this regard. One can hear a richer, more detailed, fuller mix than it's possible to capture on a stereo disc at a well-produced show. The reason for this is that a lot of gear is playing things back first-generation, directly into the P.A. mixer; the other is that the group can adapt the mix to the acoustics of the venue. Most important, the instruments can be distributed to a large number of speakers, thus making the individual voices stand out more clearly.

### Remix

The roots of remixing are in Jamaica and the dub style that evolved there in the sixties. Seminal remixer Osbourne "King Tubby" Reddock, originally a tinkerer who enjoyed fixing radios and building amplifiers, soon discovered that he had a special talent: making "specials," or one-off custom acetates, intended for use on big sound systems. "By cutting out most of the vocal track, fading it in at suitable points, reducing the mix down to the bass only, and dropping other instrumental tracks in or out, Tubby invented dub" (BBCi). Reggae producer Lee "Scratch" Perry (the Upsetters, Bob Marley and the Wailers) was also noted for his dub mixing skills, and worked with Tubby. One of Perry's signature moves was to recycle backing tracks time and again, substituting singers and renaming the single with each new iteration. Techniques pioneered in the Jamaica of the early sixties, such as dropping tracks in or out, of delay, and of swapping out vocals and backing tracks, are still in wide use today.

In the late seventies, American DJs started dissecting the three-minute song and reassembling it into an artifact suitable for dance-club use.

Remixing for extended play began in New York, in conjunction with disco music. Extended play was initially achieved by deft cutting and splicing of studio tape mixes. Early remixers include Tom Moulton as well as Larry Levan, both active in New York from the late seventies. Levan is known for DJing at the Paradise Garage on King Street (hence the genre of dance music known as "garage"), while Moulton brought remixing to the disco dance floor by creating tapes for Fire Island's Sandpiper disco in the mid-seventies. These early mixes are more properly called "re-edits," since they did not involve rebalancing and replacing elements of the mix (access to the studio multitrack masters is needed for this), but rather skillful tape editing and splicing. Moulton's rationale for his early work was that

Of course all the songs [at the disco] were three minutes long and I went "it's a shame because the minute the song is over they start mixing in this other song and they don't know whether they should dance to the new song or keep dancing to the old one."

And then people would just walk off the floor. [ . . . ] I said "there's got to be a way to make it longer where you don't lose that feeling. Where you can take them to another level." And that's when I came up with the idea to make a tape. So that's what I did. (Discoguy)

Tom Moulton is also credited with inventing the 12" single concept. In an interview, he gives a convincing account concerning how he was at a mastering facility one weekend, working with mastering engineer Jose Rodriguez, when they found that there was no more 7" stock; therefore they used 10", and found it sounded so good that they soon graduated to 12" (ibid., 2).

Once again, technology plays a role in these musical developments. As we have seen in the context of Motown, transistor radios accented midrange frequencies. Yet large disco systems featured the capacity to project low frequencies at high volume. Thus, though it may be true that the EP (extended play 12" single) was initially used by temporary necessity, the reason it came into wide use is that because of its size, the 7" single could accommodate neither longer play times, nor the wider grooves needed to cut enhanced bass frequencies into vinyl.

According to Tim Lawrence's informative study about dance music culture in the seventies, *Love Saves the Day*, Walter Gibbons was the first commercial remixer in the U.S. This is because Salsoul co-owner Ken Cayre entrusted him not only with the half-inch mixed master, but also with the actual two-inch multitrack tape of Loleatta Holloway's "Hit and Run," which Gibbons doubled in length from six to twelve minutes, turning the track into a dance floor hit.

By the late seventies, remixing began to be noticed by the pop mainstream. The actual beginning of dance remixing of rock acts is hard to trace. Rod Stewart recorded his dance hit "Do You Think I'm Sexy" in 1979, a striking stylistic departure for the rock singer. The best reliable information on the emergence of rock songs remixed for the dance floor is to be gleaned from news accounts of the early eighties, such as this 1983 excerpt from a report in *Variety*, entitled "Pop Artists Getting into Dance Music" (by Richard Gold):

New production values spawned by post-disco, modern dance music are now making an impact on mainstream rock. Some pop artists are making dance floor remixes of rock album cuts. [ . . . ] Among the established pop artists who have recently released or plan to release dance remixes are Greg Kihn, Cheap Trick, Daryl Hall and John Oates, Devo, Ric Ocasek of the Cars, (and) the Clash. [ . . . ] RCA Records black music a&r veep Robert Wright, who has made hit dance remixes for Hall & Oates, pointed to a new "fusion of cultures" that's striking a balance between black radio's emphasis on "beat and rhythm" and pop radio's orientation toward "melody and lyrics."

Initially, the dance remix was just a way to get songs onto the dance floor. Yet the DJ was also becoming an artist: at times, a remix could supplant the original producer's work. Gradually, the DJ became a producer in his own right: the ability to reorganize, to reimagine and recontextualize tracks evolved through a synergy of technology and imagination. Remixers have performed sonic alchemy on countless tracks. In the process of remaking songs, they have also crossed cultural boundaries by resituating the source material's context into a variety of dance and abstract music genres. Producer Trevor Horn's work with Frankie Goes to Hollywood is a case in point. Peter Wicke writes that Horn "produced a total of seventeen different versions of 'Relax,' which in some cases differed quite substantially, and then released these one after the other" (Wicke, 15).

Discussing the concept of the the emergence of the DJ as a creative force though creative use of the studio, Kai Fikentscher writes:

From its inception in the mid-Seventies, the 12-inch single became a vehicle for the emancipation of the disk-jockey, documenting his development from a more passive role as turntable operator to more active ones in the fields of song writing, production and engineering. Via the 12-inch single, many a DJ has risen to the status of artist, and this has necessitated a redefinition of such familiar concepts as musical instrument, performer and the role of audience in performance. (Fikentscher, 52)

The stylistic range that remixers generally work within is a genre of dance music such as house (hard, acid, etc.), jungle, or techno. Each of these styles is defined by its approach to both timbre (achieved by very specific synthesizers: e.g., Roland TB-303) and rhythm. Techno relies on the 808 beat box for its drum sound and feel, while jungle favors scratchy drum-break snippets of sixties funk such as James Brown's "Funky Drummer." In recent times, remixers have become increasingly virtuosic, harnessing studio technology to not only rebalance or extend but practically rewrite some very familiar tracks.

### "Break on Through" (The Doors; BT Remix)

This remix is captivating from the standpoints of sound, marketing, and conception. Sonically, it offers an accomplished update of the Doors classic. It's a remake of an anarchic and energetic sixties rock hit. The remix keeps the original vocals and discards everything else, replacing and rethinking the original instrumental performances entirely. The new tracks are tight, disciplined, and captivating, placing Jim Morrison's voice in a completely new context without taking away any of his passionate vocals and poetic imagery. From a marketing standpoint, it's interesting that this song is currently available only online, and that it is to appear as part of a game from Electronic Arts, "FIFA Street." Conceptually, it's interesting that Brian Transeau (BT) is listed as remixer and producer, while the surviving three members of the Doors, Ray Manzarek, John Densmore, and Robby Krieger, are listed as coproducers. It's as though he has taken over Jim Morrison's central position in the band.

The original recording quickly builds a momentum that seems to take off in every direction at once. The song opens with a bossa nova beat by drummer John Densmore, joined almost immediately by a growling electric piano and electric guitar. The ear searches in vain for the expected bass guitar, because the Doors never had a bass player (part of the reason for their quirky sound). Once the distinctive voice enters, it's clear something special is going on; poet–performer Morrison is delivering a message, both through his lyrics and his performance.

The drumming in the chorus switches to a standard rock beat featuring a very emphasized closed hi-hat, while Manzarek switches from electric piano to organ, his left hand outlining a seventh chord as a bass line while his right plays chords for emphasis. At 2:29, the original "Break on Through" is about a third as long as the remix, which clocks in at 7:07. The 1967 version has a quirky structure, too. The verse lasts all of two bars, then there's one bar of pre-chorus, followed by three bars of chorus, and the cycle begins again. This happens three times, interrupted by two instrumental breaks, and then the music is over.

The remixed "Break on Through" seeks transcendence through repetition. The track opens with drums, as does the original; but this is not a sixties drum sound, nor is it a bossa nova. Instead, we get a minute and a half of powerful, yet restrained (because compressed and gated) computer-driven drums, mixed with various bleeps, scrapes, reverse-decay cymbals, and a minimal, slowly building bass line. Morrison's voice first enters briefly at 0:26, but it is only momentary, and it is modulated so as to sound grainy, as well as being put through a delay. As he says "break on through," a lower frequency is added to the kick drum, and the synth bass is raised in the mix, thus intensifying the overall feel of the groove. Little rhythmic and timbral details are added and subtracted to keep the mix from sounding too repetitive, but Morrison's voice is not brought in again until around 1:30. Once his voice does enter, a second bass synth becomes more prominent in the mix, its resonant filter opening and closing for a wah-wah-like effect. Soon the bass synths seem to be having their own conversation under Morrison's voice, which is made to overlap itself, modulate, and fit exceedingly well with the groove. This latter effect is probably achieved by working in a digital workstation, and sliding words, syllables, even individual consonants and vowels back and forth so as to fit the groove. This technique, while labor-intensive, can be used to create the impression that the singer was listening to the remixed backing track when he sang the original.

At 3:24, the beat stops and there is a classic "breakdown," in which the synthesizers are worked against swooshing cymbals, delayed fragments of Morrison's voice, and each other. At 4:47, the groove and the hook return, with Morrison singing "break on through to the other side" again and again. Care has been taken not to fall into too much literal repetition: the fine line between the delight of recognition and potential boredom is respected. By 6:15 the voice is out, and the beat, together with the synths, is left to carry the remix out to its conclusion at 7:07.

The master remixer is himself remixed on his four-track release *Somnambulist*. The first track is his own, followed by three variations by Junkie XI, Sander Kleinenberg, and Burufunk. Each of the remixes has a distinctly different take on the original, which is already a dance track. Junkie XI's mix is house-oriented, with a heavy four-on-the-floor throughout, while Kleinenberg offers a less bassy interpretation of the same theme. Burufunk's has a more syncopated, funk/hip-hop feel.

**Re-editing Updated: Mash-Ups**

Another area of contemporary remixing seems to have returned to the era of re-edits, with a digital twist. The return to re-editing is born out of the same necessity as existed when Tom Moulton re-edited Double Exposure's "Ten Percent." It simply is not possible for the average person to get his hands on studio masters. However, the computer has introduced a new phenomenon: the mash-up. Thanks to beat, key, and tempo matching, now commonly available in inexpensive, even free music software packages, it is possible for anyone to match up (mash-up) different albums and hear how they sound together. The most notable mash-up in recent memory is Danger Mouse's *Grey Album*, which famously mixes together the Beatles' *White Album* and Jay-Z's *Black Album*. Danger Mouse's approach was to loop short segments of Beatles tracks behind Jay-Z's voice. The track "99 Reasons" works remarkably well, mashing together segments of the Beatles' "Helter Skelter" and Jay-Z's rap.

There is even a Jay-Z Construction Set, 650 MB of samples and other remixes, available via P2P (peer-to-peer) networking. The set is distributed using a method adapted from Napster: a P2P program called Bit Torrent links computers that have already downloaded the Construction Set to those that wish to do so, thus decentralizing the data-intensive downloads. Once again, technology makes new forms of music possible—or perhaps the new music springs up as a result of the technology being available.

Once the prospective remixer has obtained the Construction Set, he can listen to what others have done with the samples, and construct his own, uploading it onto the Internet for all to hear. By employing the four-square computerized grid of the digital sequencer as his principal organizing device, the remixer acts at once as a reductionist, a leveler of all musics, but also as someone who crosses cultural boundaries. The glue between styles is the four-on-the-floor beat (bass drum every quarter note): the most diverse types of music can be (and are) made to fit together over the common denominator of a steady beat.

Given the times in the past that a new technology has failed to usher in a utopian age of music, one is hesitant to make pronouncements. But it does seem as though there is a more participatory entertainment culture afoot. As one message board participant put it, "Remember TV? It was like the Internet, only you couldn't do anything with it."

## The Contemporary Situation: Is the Producer Obsolete?

In the course of looking at the way the role of the producer has changed over time, we have watched it expand consistently. The places one may expect to find the producer have become more and more inclusive, ranging from sitting at the back of the control room (the fifties), to moving up to the board (the sixties), to taking over the job of the composer (the seventies and eighties), to becoming the performer onstage (the nineties).

The process has been additive, so that today one can find producers in any one or all of these functions. In some ways, it's more difficult than ever to pin down what the role of the producer is. This is partly a result of the explosion of a great variety of recorded musics, as well as the techniques which have been developed to create and capture them.

From the beginning, the producer was a point man, a kind of technical-artistic interpreter. Fred Gaisberg knew the proper distance for Enrico Caruso to stand from the gramophone needle as it etched his voice into acetate. In time, the matrix of recording was reversed, and the idea of reproducing the concert hall experience had to make room for another aesthetic: the ideal of creating an illusion of reality on recordings was joined, and to some extent supplanted by, the reality of the illusions themselves. As of the sixties, the new way of recording was firmly established, having moved from novelty status to central prominence on albums such as *Sergeant Pepper*.

But the old ways did not just disappear. There are many important producers today creating music which does seek to convey the sound of a live band playing real instruments; in rock, Steve Albini and Butch Vig are two prominent examples. What has happened is that the horizons of recording, indeed of music as a whole have expanded. As new production technologies and aesthetics have emerged, the old techniques have not simply fallen away. While it is true that most everyone has embraced a certain amount of change (I'm not aware of anyone besides Phil Spector championing mono, for example), there is an enormous range of recording philosophies in the world. It would be foolish to claim that the old ways are obsolete. In a sense, there are as many styles of production as there are producers.

Much as the term "computer music" has ceased to have any stylistic connotation owing to the ubiquity of the microprocessor, it may not make much sense any more to speak of "the producer" as a clearly delineated entity. He (or *she*, one hopes; the field is still almost entirely male-dominated) can play many roles, even varying from project to project.

The reason for the sharp rise in the number of producers in the world today is music technology and its interdependence with pop music. Both are constantly evolving and influencing each other. Consider the example of the drum machine: created at first to replace the drummer as realistically as possible, it quickly spawned music which embraced and built on its mechanical feel. Disco, hip-hop, and electronica all owe their robotic, regular grooves to the robtic internal clocks of Linn Drums and Roland 808s. People were needed who could work such machines; those who did so in a creative way were eventually called producers. Other technologies have had a similar effect. New skills are necessary to master new technologies, and those new technologies are used in ways often not foreseen by their inventors.

Anyone who has so much as arranged loops in a freeware program or made music with a MIDI sequencer has learned something about production: balances, choice of sounds, and methods of recording and mixing are all elements of the art. In popular music, production skills have already joined (and in some cases supplanted) keyboard skills as part of the *lingua franca* of music. In hip-hop, for example, the quality of the beat and the rap is far more important than the concept of a key, let alone modulation.

Simon Reynolds has interesting things to say about the role of the producer in the context of dance music:

House makes the producer, not the singer, the star. It's the culmination of an unwritten (because unwritable) history of black dance pop, a history determined not by sacred cow auteurs but by producers, session musicians, and engineers—backroom boys. House music takes this depersonalization further: it gets rid of human musicians (the house band that gave Motown or Stax or Studio One its distinctive sound), leaving just the producer and his machines. [ . . . ] Closer to an architect or draftsman, the house auteur is absent from his own creation; house tracks are less like artworks, in the expressive sense, than vehicles, rhythmic engines that take the dancer on a ride. (Reynolds, 30)

This statement implies an interesting problem. While it may have been true that the house producer was not generally a performer in the eighties, we have seen that artist–engineer–producers such as Outkast, the Chemical Brothers, Daft Punk, and many others are now taking their studios onto the stage. So it seems that what has happened, at least in the world of electronic pop, is that the role of the producer has now merged with that of the engineer and performer. Technology is what has made this possible: Andre' 3000 is imagining his sounds in terms of machines, and he knows how to get those sounds because they are readily accessible; moreover, modern recording technology obviates manual dexterity. What is important now is imagination.

In discussing recording and production, Reynolds and other writers have an egalitarian agenda at heart, which is commendable. But it seems that there will always be people who do something particularly well, no matter whether the context is marks on paper or samples in the studio. Performance is not likely to disappear anytime soon, as it fulfills a basic human need. It's not surprising that far from removing "sacred cow auteurs," modern technology has simply shifted the metaphor from exceptional accomplishment on paper by "composers" to exceptional accomplishment on hard disk by "producers." Moreover, the producer and his machines are on stage, just as the composer was once a performer. At the top of the current charts, one increasingly finds cases in which the producer is the artist is the composer is the producer; and technology is what has driven the change.

# Glossary

Terms in italics have their own entries.

**acetate**  In early recording, sound grooves were etched onto a disc of cellulose acetate. The sound quality was high, but the discs wore out after only a few playings. Generally used for test pressings, and for masters from which vinyl LPs were made.

**active pickup**  Found on electric guitars and basses. Active electronics in the *pickup* module permit more extensive shaping on the instrument itself of the output sound.

**ADC**  Analog to digital conversion.

**amp**  An abbreviation for amplifier. A circuit designed to increase the level of an incoming signal.

**analog**  Characterized by continuous waves, as opposed to the binary world of *digital*. (The term stems from the creation of electrical signals that are *analogous* to audible fluctuations in sound pressure.)

**backline**  Amplifiers, drums, and other equipment found on stage.

**bandpass**  A type of *filter* that allows only a limited *frequency range* to pass through it.

**capstan**  In a tape recorder, a motor-driven, rotating post that, with the help of the *pinch-roller,* contacts the tape and pulls it through at a constant speed.

**cardioid**  A heart-shaped *polar pattern* found on many *microphones.*

**CD**  Compact disc.

**chorus**  An audio *effect* in which a source signal is time-delayed, detuned, and mixed with the unprocessed signal in order to create the illusion of multiple sources.

**click track**  An electronic metronome, usually recorded onto a track and fed to the musician's headphones as a timing reference.

**compression**  Dynamic range limiting applied to audio signals. Used to decrease the peaks of a signal to allow for amplification without distortion, as well as to change the timbral characteristics of a sound by altering its dynamics.

**condenser**   A *microphone* that uses the principle of variable capacitance to translate acoustic sound pressure into an electrical signal. Condenser (or capacitor) microphones require a power supply, typically called phantom power, to operate.

**cover**   An existing song, performed or recorded by another musician or group.

**crossover (style + electronic)**   Style: a song that blends different genres, or a song that's in a genre that the musician or group is not normally associated with. **Electronic:** an electronic circuit consisting of *filters* which split an audio signal into different frequency bands, allowing the filtered signals to be routed to speakers designed for a particular *frequency range*.

**DAC**   *Digital* to *analog* conversion.

**DAW**   *Digital* audio workstation. An integrated recording, editing, mixing, and mastering environment.

**decibel**   A logarithmic measurement of sound pressure or voltage level.

**delay**   An audio *effect* that replays an input signal after a specified period of time. Often delays provide a feedback control in order to send the delayed signal (the output) back to the input, creating multiple echoes of the source sound.

**digital**   Information stored as a series of binary numbers. Digital audio stores the instantaneous amplitudes of a waveform after it has gone through the process of analog-to-digital (*ADC*) conversion.

**distortion**   In general, any difference between an original signal and one that has been processed, aside from relative level. Typically distortion refers to the overamplification of an audio signal resulting in undesired effects.

**ducking**   In recording, the use of the signal of one channel to control the volume of one or more other channels. Used to foreground the voice.

**effects**   Processes applied to audio signals to alter their sonic characteristics.

**EQ (equalizer)**   A device employing multiple *filters*; used to alter the frequency content of an audio signal.

**equalization (EQing)**   The process of using an equalizer on an audio signal.

**erase**   On an *analog* tape machine, the erase head randomizes the tiny magnets on the tape so the encoded pattern disappears and the audio is lost. On computers or *digital* equipment, erasing is essentially freeing the memory location for new information.

**fader**   A type of volume control usually associated with audio *mixers*.

**feedback**   Any situation where the output of a system is able to reenter the system. In extreme situations, this can cause the system itself to resonate at certain frequencies, often creating loud, pitched sounds.

**fft**   Fast Fourier Transform. In the 18th century the French mathematician Fourier postulated that all sounds could be reduced to an (infinite) set of sine waves. In general, the Fast Fourier Transform provides a shift from the time-domain, where sounds are analyzed as amplitude fluctuations over time, to the frequency-domain, where sounds are analyzed by their frequency content over time.

**figure (of) eight (bipolar)**   A *microphone polar pattern* that provides equal sensitivity in front and in back of the microphone. The response is least sensitive 90 degrees off axis.

**filter**   A process that can be applied to an audio signal in order to alter the frequency content of the signal.

**5.1**   Primary format for *surround sound*. The numbers denote channels: five principal channels, and one for LFE (low frequency effects). This is currently the standard used on *DVD-A* and *DVD-V* discs.

**flanging**   An audio effect in which a source signal is time-delayed, filtered, and mixed with the unprocessed signal in order to create *phase interference*.

**Fletcher-Munson curve**   A perceived loudness curve that shows that humans are less sensitive to low and high frequencies, and more sensitive to middle-range frequencies at low volumes.

**footpedal**   A piece of signal-processing equipment housed in a small case with a switch on top to activate and deactivate the processing. Also known as a stompbox.

**frequency range**   A specified range of operation. The human ear, for example, is capable of perceiving roughly the 20Hz–20kHz range. Thus most *microphones* and loudspeakers are engineered for a similar range.

**fx**   Shorthand for *effects*.

**high-pass**   A *filter* that allows frequencies above a (selectable) frequency to pass, removing the low frequency content from the signal.

**hiss**   Small amounts of randomization in the magnets on *analog* tape that become audible when amplified for output to speakers.

**low-pass**   A *filter* that allows frequencies below a (selectable) frequency to pass, removing the high-frequency content from the signal.

**mic**   Abbreviation for *microphone*.

**microphone**   A piece of electronic equipment that transfers sound pressure into an analogous electrical voltage. The primary styles of microphone are dynamic, *condenser*, and *ribbon*.

**mix**   The product of the *mixdown*.

**mixdown**   The process of combining recorded *tracks* into a *mix*, usually by copying multiple tracks down to two tracks. During this process, volumes are adjusted, *EQ* is tweaked, and final *effects* (e.g. *reverb*) are applied.

**mixer, mixing board**   An electrical device used to receive, balance, process, and output audio signals.

**multitrack**   A recording that contains multiple signals recorded onto multiple tracks in order to allow for detailed mixing at a later point.

**native processing**   Computer processing handled by a machine's built-in CPU, as opposed to an outside unit or an add-on card.

**noise**   A randomized signal with no periodicity. White noise contains equal power per frequency. Pink noise contains equal power per octave.

**nonlinear editing**   The use of computers in sound editing, which allows copying and pasting sound without signal degradation. Replaces tape *splicing.*

**omnidirectional**   The *polar pattern* of a *microphone* that responds equally to sounds from any direction.

**outboard**   Processing equipment outside of the *mixing board* or computer. Typically associated with compressors, *equalizers*, and *preamps* as these are often found in an audio *mixer.*

**overdrive**   Amplification of a signal to the point of distortion to achieve a desired effect.

**overdub**   To record new *tracks* on a *multitrack* recording in synchronization with previously recorded tracks.

**parametric**   Generally applied to *EQ*. Denotes added parameters: where a standard EQ allows adding or subtracting a fixed frequency band, parametrics also allow user selection of the band, as well as of its curve (from narrow, affecting only a few frequencies, to wide, affecting many).

**pickup**   A device housed in an instrument to convert sound pressure to electrical signals.

**phase interference**   When two signals are added together, the peaks and dips in amplitude are added together. If the signals have similar waveforms but are offset in time, the dips in one signal may add together with the peaks of the other, resulting in phase cancellation. If the waveforms are similar enough, any time offset will result in audible filtering of frequency content and alterations in special imaging over loudspeakers.

**pinch-roller**   In a tape recorder, a freely turning wheel that keeps the tape flush against the *capstan* to allow the tape to be pulled through at a constant speed.

**playback**   When a song or *track* is played after it has been recorded.

**plug-in**  An individual piece of software written to perform signal processing and used in *DAW*s.

**polarity**  A condition of having two states, typically represented as + and – in audio equipment to represent an increase or decrease in voltage.

**polar pattern**  A plot of a device's sensitivity as a function of the angle around the device. Often associated with *microphones*, which can exhibit *omnidirectional, cardioid,* and *figure-eight* patterns.

**postproduction**  The process of enhancing production values in the studio by editing and adding elements. This includes ADR, *effects*, remix, etc. Used primarily in a sound-for-picture context.

**preamp**  Abbreviation for preamplifier: prepares the signal for output to a power amplifier. (Often associated with *microphones*. Allows a variety of microphone output levels to be amplified to more appropriate levels for use in other audio equipment, such as *mixers* or recording devices.)

**preproduction**  The process prior to the creation of music or sound, primarily in a sound-for-picture context. This can include finding musicians, composers, conductors, creating a budget or a list of equipment needs, etc.

**processor channel strip**  A sequence of *effect* processing for one signal, housed in a single unit for convenience. A channel strip often provides amplification, *equalization,* and *compression* capabilities.

**project studio**  A small studio built specifically for certain commercial markets.

**punching in**  Replacing a (generally short) segment of a recording with a new take. The *record* button must be pressed at exactly the right moment.

**record**  1. *Analog* media that contains grooves cut into vinyl. 2. To store sonic information on a specified recording medium.

**reverb**  Abbreviation for reverberation. Either the natural characteristics of sound reflections in an acoustic space, or processing designed to simulate these natural characteristics.

**shelving**  A *filter* that can attenuate either high or low frequencies by a specified volume without completely eliminating them, as opposed to a *low-pass* or *high-pass* filter.

**splicing**  The cutting of magnetic tape, usually in the context of editing.

**stompbox**  Another term for a *footpedal*.

**studio tan**  Acquired by those who spend too much time in the studio, especially tweaking and fixing minor details in the *mix*.

**surround**   Abbreviation for surround sound. Sound occurring from all sides of a listener, as opposed to stereo in which the sound field is contained in front of the listener.

**tape head**   A tape machine usually consists of three tape heads: the *record* head, which converts electrical signals into an analogous magnetic pattern stored on the tape; the *playback* head, which reads the magnetic pattern stored on the tape and converts it into an electrical signal; and the *erase* head, which purposely randomizes the small magnets on a tape to erase any previously recorded signals.

**track**   A designated space on a recording medium designed to store one signal.

**tracking**   The process of recording instruments to a recording medium.

**transformer**   A device used to transfer electrical energy from one circuit to another.

**vintage**   A technique, or especially a piece of equipment, of a certain age. Generally pre-1985, e.g., the Fender Rhodes electric piano.

**virtual studio**   A representation of a recording studio inside a computer. The *mixer*, tape recorder, *effects*, synthesizers, and even acoustic instruments are increasingly well modeled in software that will run on a computer.

**volume envelope**   The loudness contour of a sound over time. A piano's volume envelope, for example, starts off with a quick peak and then decays slowly (if pedal down).

**VU meter**   From "volume unit" meter. Originally, an audio output meter for broadcast and recording studio consoles with defined reference levels and movement characteristics. It has become a common name for most audio output meters.

**world-class studio**   As opposed to a *project studio* or home studio; a recording studio that contains extremely high-end equipment, one or more pleasing acoustic environments to *record* in, a controlled acoustic environment to *mix* in, and knowledgeable staff engineers.

# Recordings Cited

Aerosmith. "Walk This Way." *Toys in the Attic.* 1975. Sony/Columbia, 1993.

Baker, Ginger. *Horses and Trees.* 1986. Charly, 1998.

Bambataa, Afrika. "Planet Rock." *Street Jams: Electric Funk.*

The Beach Boys. *20 Good Vibrations: The Greatest Hits.* Capitol, 1995.

——. "Good Vibrations." *20 Good Vibrations.* Capitol, 1995.

——. "I Get Around." *20 Good Vibrations.* Capitol, 1995.

——. "Help Me Rhonda." *20 Good Vibrations.* Capitol, 1995.

——. "Surfin' U.S.A." *20 Good Vibrations.* Capitol, 1995.

——. *The Making of "Pet Sounds."* 4-CD set. Capitol, 1996.

The Beastie Boys. "Fight for Your Right to Party." *Licensed to Ill.* Def Jam, 1986.

The Beatles. *Abbey Road.* 1969. Capitol/EMI, 1987.

——. "Love Me Do." 1963. *The Beatles: 1962–1966.* Capitol/EMI, 1993.

——. *Magical Mystery Tour.* 1967. Capitol/EMI, 1987.

——. "Please Please Me." 1963. *The Beatles: 1962–1966.* Capitol/EMI, 1988.

——. *Revolver.* 1966. Capitol/EMI, n.d.

——. *Sergeant Pepper's Lonely Hearts Club Band.* 1967. Capitol/EMI, 1987.

——. "Strawberry Fields." 1967. *Magical Mystery Tour.* Capitol/EMI, 1987.

——. "Taxman." *Revolver.* 1966. Capitol/EMI, n.d.

——. *The White Album.* 1968. Capitol/EMI, 1987.

Beck. *Odelay.* Geffen, 1996.

Big Mama Thornton. "Hound Dog." 1952. *Hound Dog: The Peacock Recordings.* Uni/MCA, 1992.

Bowie, David. *The Rise and Fall of Ziggie Stardust*. 1972. Virgin, 1999.

———. *Heroes*. 1977. Rykodisc, 1991.

———. *Low*. 1977. Virgin, 1999.

Brown, James. "The Funky Drummer." *Star Time*. 4-CD set. Uni/Mercury, 1991.

The Chemical Brothers. *Exit Planet Dust*. Astralwerks, 1994.

———. *Dig Your Own Hole*. Astralwerks, 1997.

———. *Surrender*. Astralwerks, 1999.

The Crystals. "Uptown." On Spector, *Back to Mono*, CD 2.

Daft Punk. "Daftendirekt." *Homework*. Virgin, 1996.

———. *Homework*. Virgin, 1996.

Danger Mouse. *The Grey Album*. November 29, 2004. See http://www.illegal-art.org/audio/grey.html.

Dibango, Mano. *Afrijazzy*. Enemy, 1994.

The Drifters. *Definitive Drifters Anthology Volume Three: Save the Last Dance for Me*. Sequel, 1996.

Eno, Brian. *Discreet Music*. 1975. Editions EG, 1989.

———. *Here Come the Warm Jets*. 1973. Editions EG, 1989.

———. *Music for Airports*. 1978. Editions EG, 1989.

———. *Music for Films*. 1978. Editions EG, 1989.

———. (Prod.) *No New York*. Various artists. Sire, 1978.

———, and David Byrne. *My Life in the Bush of Ghosts*. Sire, 1981.

Goldie. *Timeless*. ffrr, 1995.

Grandmaster Flash. *The Message*. Sugarhill, 1982.

———. "White Lines." *White Lines*. 1981. Hot Records, 1997.

Hancock, Herbie. "Rockit." *Street Jams: Electric Funk*. Rhino, 1992.

Hendrix, Jimi. *Axis: Bold as Love*. 1967. Uni/MCA, 1997.

Jagger, Mick. *She's the Boss*. 1984. WEA/Atlantic, 1993.

Jackson, Michael. *Off the Wall: Special Edition*. Epic, 2001.

———. *Thriller: Special Edition*. Epic, 2001.

Joy Division. *Permanent: Joy Division 1995*. Warner, 1995.

Jay-Z. *The Black Album*. Roc-A-Fella, 2003.

Kraftwerk. "Numbers." *Computer World*. 1981. WEA/Elektra, 1988.

————. *Trans-Europe Express*. 1976. Capitol/EMI, 1995.

Led Zeppelin. "Whole Lotta Love." *Led Zeppelin II*. 1969. Atlantic, 1990.

————. *Led Zeppelin IV*. 1973. Atlantic, 1990.

"Linn Drum." June 2000. See www.ths.com/linndrum.

Manson, Marilyn. *Antichrist Superstar*. Nothing/Interscope, 1997.

Moby. *Play*. V2/BMG, 1999.

Moulton, Tom, et al. *Salsoul Presents: The Definitive 12" Masters, vol. 1*. Salsoul U.K/Reisuue, 2003.

Nine Inch Nails. *The Downward Spiral*. Nothing/Interscope, 1994.

Peterson, Ray. "*Corrine, Corrina*." On Spector, *Back to Mono*, CD 1.

Pink Floyd. *Atom Heart Mother*. 1970. Capitol/EMI, 1994.

————. *Dark Side of the Moon*. 1973. Capitol/EMI, 1994.

————. *Meddle*. 1971. Capitol/EMI, 1994.

————. "One of These Days." *Meddle*. Capitol/EMI, 1994.

————. *Piper at the Gates of Dawn*. 1967. Capitol/EMI, 1987.

————. *Ummagumma*. 1969. Capitol/EMI, 1987.

Pop, Iggy. *Lust for Life*. 1977. Virgin, 1990.

Prong. *Third Planet from the Sun*. EP. Southern, 1987.

Public Enemy. *It Takes a Nation of Millions to Hold Us Back*. Def Jam, 1988.

Public Image Limited. *Album*. Island, 1987.

The Ramones. "Rock'n'Roll High School!" *End of the Century*. 1980. Warner, 1994.

Ribovsky, Mark. *He's a Rebel*. New York: E. P. Dutton, 1989.

The Righteous Brothers. "You've Lost That Lovin' Feelin'." On Spector, *Back to Mono*, CD 3.

Riley, Terry. *Rainbow in Curved Air*. Columbia, 1967.

The Ronettes. "Be My Baby." On Spector, *Back to Mono*, CD 2.

Sly and the Family Stone. *Greatest Hits*. Epic/CBS, n.d.

Spector, Phil. (Prod.). *Back to Mono: (1958–1969)*. 4-CD set. Phil Spector Records, 1991.

———. *A Christmas Album. Back to Mono*, CD 4.

Squarepusher. *Bigloada*. Nothing/Warp, 1997.

*Street Jams: Electric Funk Part 1*. Rhino, 1992.

*Street Jams Sampler*. Rhino, 1992.

Summer, Donna. *On the Radio: Greatest Hits Volume I and II*. Casablanca, 1979.

———. "I Feel Love." 1977. *On the Radio*. Casablanca, 1979.

———. "Love to Love You Baby." 1975. *On The Radio*. Casablanca, 1979.

The Sugarhill Gang. "Rapper's Delight." *Street Jams Sampler*. Rhino, 1992.

Swans. *The Burning World*. Uni/MCA, 1989.

———. *Children of God*. Caroline, 1987.

———. *Cop*. 1984. Thirsty Ear, 1999.

———. *Greed*. 1985. Thirsty Ear, 1999.

———. *Love of Life*. Young God, 1992.

———. "See No More." Preproduction click track and live version. September 15, 2000. See http://www.virgilmoorefield.com/producersounds.html.

———. "See No More." *The Burning World*. Uni/MCA, 1989.

———. *Various Failures: 1988–1992*. Young God, 1998.

———. *White Light from the Mouth of Infinity*. Young God, 1991.

The Teddy Bears. "To Know Him Is to Love Him." Spector, CD 1.

Turner, Ike, and Tina Turner. "River Deep, Mountain High." On Spector, *Back to Mono*, CD 3.

U2. *Zooropa*. Island, 1993.

The Velvet Underground. *White Light, White Heat*. 1968. Polydor, n.d.

The Who. "Baba O'Reilly." *Who's Next*. MCA, 1995.

———. *Tommy*. 1970. MCA, 1995.

————. *Who's Next*. 1971. MCA, 1995.

Zappa, Frank. *Freak Out*. Verve, 1967.

————. *Hot Rats*. Warner/Reprise, 1969.

————. *Lumpy Gravy*. Warner/Reprise, 1966.

————. *Uncle Meat*. Verve, 1968.

————. *We're Only in It for the Money*. Warner/Reprise, 1967.

————. *We're Only in It for the Money*. Remixed and remastered. Rykodisc, 1986.

# Bibliography

Abbott, Kingsley. *The Beach Boys' "Pet Sounds": The Greatest Album of the Twentieth Century*. London: Helter Skelter Publishing, 2001.

Attali, Jacques. *Noise: The Political Economy of Music*. Trans. B. Massumi. Minneapolis: University of Minnesota Press, 1985.

Bessman, Jim. *Ramones: An American Band*. New York: St. Martin's Press, 1993.

"Bill Laswell." Accessed August 9, 2000, at http://www.hyperreal.org/music/labels/axiom.

Brackett, David. *Interpreting Popular Music*. Cambridge: Cambridge University Press, 1995.

Brewster, Bill, and Frank Broughton. *Last Night a DJ Saved My Life: The History of the Disc Jockey*. New York: Grove Press, 2000.

Burgess, Richard James. *The Art of Record Production*. London: Omnibus Press, 1997.

Bussy, Pascal. *Kraftwek: Man, Machine, and Music*. London: SAF Publishing, 1993.

"Chemical Brothers." Accessed September 9, 2000, at http://the-raft.com/chemicalbros/biography.html.

Clark, Martin. "Working the Web: Dance Music." *Guardian,* October 11, 2001. Accessed February 24, 2005 at http://www.guardian.co.uk/internetnews/story0,7369,567111,00.html.

Cook, Perry, and Dan Levitin. "Memory for Musical Tempo: Additional Evidence That Auditory Memory Is Absolute." *Perception and Psychophysics* 58.6 (1996): 927–935.

Courrier, Kevin. *Dangerous Kitchen: The Subversive World of Frank Zappa*. Toronto: ECW Press, 2002.

Cunningham, Mark. *Good Vibrations: A History of Record Production*. London: Sanctuary Publishing, 1998.

Curtis, Anthony, and James Henke, eds. *The Rolling Stone History of Rock and Roll.* New York: Random House, 1992.

Discoguy. "Tom Moulton." Accessed February 24, 2005 at http://www.disco-disco.com/tributes/tom.html.

Eisenberg, Evan. *The Recording Angel: Explorations in Phonography.* New York: McGraw-Hill, 1987.

Eno, Brian. "The Studio as Compositional Tool." *Downbeat* 56–57, 50–52 (July/August 1983).

Everett, Walter. *The Beatles as Musicians.* 2 vols. New York: Oxford University Press, 2001.

Fikentscher, Kai. *"You Better Work!": Underground Dance Music in New York City.* Hanover, N.H.: Wesleyan Press, 2000.

Fitzgerald, John. "Motown Crossover Hits 1963–1966 and the Creative Process." *Popular Music* 14.1 (1995): 1–11.

Fox, Ted. *In the Groove: The People behind the Music.* New York: St. Martin's Press, 1986.

George, Nelson. "Standing in the Shadows of Motown." *Musician* 60 (1983): 61–66.

Gerrish, Bruce. *Remix: The Electronic Music Explosion.* Vallejo, Calif.: Artistpro, 2001.

Goehr, Lydia. *The Imaginary Museum of Musical Works.* Oxford: Clarendon Press, 1992.

Gold, Richard. "Pop Artists Getting into Dance Music." *Variety* 310.12 (March 9, 1983).

Goldman, Albert. *Disco.* New York: Hawthorn Books, 1978.

Goldberg, M., and N. Seeff. "Berry Gordy." *Rolling Stone* 585 (1990): 66–72.

Gronow, Pekka, and Ilpo Saunio. *An International History of the Recording Industry.* London and New York: Cassell, 1998.

Haa, Erika. *Boogie Nights: The Disco Age.* New York: Friedman/Fairfax Publishers, 1994.

Haden-Guest, Anthony. *The Last Party: Studio 54, Disco, and the Culture of the Night.* New York: William Morrow, 1997.

Hager, Steven. *Hip Hop: The Illustrated History of Break Dancing, Rap Music, and Graffiti.* New York: St. Martin's Press, 1984.

Hitchcock, Wiley, ed. *The Phonograph and Our Musical Life.* New York: I.S.A.M. Monographs, 1980.

Jones, Steven. "Rock Formation: Popular Music and the Technology of Sound Recording." Ph.D. Dissertation, University of Illinois Urbana-Champaign, 1987.

"King Tubby." Accessed November 10, 2004, at http://www.culture.court.com/Br. Paul/media/KingTubby.htm.

"King Tubby." Accessed November 11, 2004, at http://www.furious.com/perfect/ KingTubby2.html.

"King Tubby." Accessed November 11, 2004 at http://www.bbc.co.uk/cgi-perl/ music/muze/index.pl?site=music&action=biography&artist_id=16477.

Kostelanetz, Richard, ed. *The Frank Zappa Companion.* New York: Schirmer Books, 1997.

Lawrence, Tim. *Love Saves the Day.* Durham, N.C.: Duke University Press, 2003.

Leaf, M. *The Making of* Pet Sounds. Capitol Records, 1996.

Lewisohn, Mark. *The Complete Beatles Recording Sessions.* London: EMI, 1989.

Lindeman, Steve. "Fix It in the Mix." *Popular Music and Society* 22.4 (1998).

Longhurst, Brian. *Popular Music and Society.* Cambridge: Polity Press, 1995.

MacDonald, Ian. *Revolution in the Head: The Beatles' Records and the Sixties.* London: Pimlico, 1995.

Martin, George, and Jeremy Hornsby. *All You Need Is Ears.* New York: St. Martin's Press, 1979.

Martin, George, and William Pearson. *Summer of Love: The Making of* Sergeant Pepper. London: Pan Books, 1995.

Massey, Howard. *Behind the Glass: Top Record Producers Tell How They Craft the Hits.* San Francisco: Backbeat Books, 2001.

Michie, Chris. "You Call That Music?" Accessed November 15, 2004, at http:// mixonline.com/recording/interviews/audio_call_music/index.html.

Michie, Chris. "We are the Mothers . . . and This Is What We Sound Like!" Accessed November 15, 2004, at http://mixonline.com/recording/business/audio_mothers_ sound/index.html.

Michie, Chris. "The Complete Mark Pinske Interview." Accessed November 15, 2004, at http://mixonline.com/mag/audio_complete_mark_pinski/index.html.

Ogg, Alex, and David Upshall. *The Hip-Hop Years: A History of Rap.* London: Channel 4 Books, 1999.

Palmer, Robert. "Leiber and Stoller: The Rock & Roll Years." In *Baby, That Was Rock & Roll*. New York and London: Harcourt Brace Jovanovich, 1978.

Palmer, Tamara. "Autechre." Accessed September 18, 2000, at http://www.warp.com/autechre.

"Performing Engineers." *EQ* (April 1999).

Posner, Gerald. *Motown: Music, Money, Sex, and Power*. New York: Random House, 2002.

Reynolds, Simon. *Generation Ecstasy: Into the World of Techno and Rave Culture*. New York: Routledge, 1999.

Ribovsky, Mark. *He's a Rebel*. New York: E. P. Dutton, 1989.

Ross, Sean. Liner notes, *Street Jams: Electric Funk Part 1*. Rhino, 1992.

Rule, Greg. "Trent Reznor." *Keyboard* (March 1994).

"The Russell Simmons Countdown of the Top Ten Producers in Hip-Hop." September 9, 2000. ABC Television Network.

Shaw, Greg. "Brill Building Pop." In *The Rolling Stone History of Rock and Roll,* ed. Anthony Curtis and James Henke. New York: Random House, 1992.

Slutsky, Alan. *Standing in the Shadows of Motown*. Motion picture (dir. Paul Justman). DVD: Artisan Entertainment, 2002.

Strauss, D. "Sound Found: Producer and Engineer Conny Plank Helped German Music Find Its Voice." *Remix* (August 2004). Accessed February 24, 2005 at http://remixmag.com/mag/remix_conny_plank.

Tamm, Eric. *Brian Eno: His Music and the Vertical Color of Sound*. New York: Da Capo Press, 1995.

Tobler, John, and Stuart Grundy. *The Record Producers*. New York: St. Martin's Press, 1982.

Unterberger, Richie. *Unknown Legends of Rock'n'Roll*. San Francisco: Miller Freeman Books, 1998.

Walser, Robert. *Running with the Devil: Power, Gender, and Madness in Heavy Metal Music*. Hanover, N.H.: Wesleyan University Press, 1993.

Warner, Timothy. *Pop Music—Technology and Creativity*. Burlington, Vermont: Ashgate, 2003.

Watson, Ben. "Frank Zappa as Dadaist: Recording Technology and the Power to Repeat." In *The Frank Zappa Companion,* ed. Richard Kostelanetz. New York: Schirmer Books, 1997.

Wicke, Peter. *Rock Music: Culture, Aesthetics, and Sociology*. Cambridge: Cambridge University Press, 1987.

Wilson, Brian. *Wouldn't It Be Nice: My Own Story*. New York: Harper Collins, 1991.

Wilson, Olly. "The Significance of the Relationship between Afro-American Music and West African Music." *The Black Perspective in Music* 2 (spring 1974): 3–22.

Wolfe, Tom. "The First Tycoon of Teen." *The Kandy-Kolored Tangerine-Flake Streamline Baby*. New York: Farrar, Straus, and Giroux, 1965.

Zak, Albin. *The Poetics of Rock: Cutting Tracks, Making Records*. Berkeley: University of California Press, 2001.

# Index